New Europe:
Imagined Spaces

New Europe:
Imagined Spaces
Donald McNeill

A member of the Hodder Headline Group
LONDON

Distributed in the United States of America by Oxford University Press Inc.,
New York

First published in Great Britain in 2004 by
Arnold, a member of the Hodder Headline Group,
338 Euston Road, London NW1 3BH

http://www.arnoldpublishers.com

Distributed in the United States of America by
Oxford University Press Inc.
198 Madison Avenue, New York, NY 10016

The advice and information in this book are believed to be true and
accurate at the date of going to press, but neither the author nor the publisher
can accept any legal responsibility or liability for any errors or omissions.

British Library Cataloguing in Publication Data
A catalogue record for this book is available from the British Library

Library of Congress Cataloging-in-Publication Data
A catalog record for this book is available from the Library of Congress

ISBN 0 340 76055 9

1 2 3 4 5 6 7 8 9 10

Typeset in 10/14pt Gill Light by Phoenix Photosetting, Chatham, Kent
Printed and bound in Great Britain by The Bath Press Ltd., Bath

What do you think about this book? Or any other Arnold title?
Please send your comments to feedback.arnold@hodder.co.uk

Contents

Conents

Acknowledgements

This book germinated in Glasgow, in a couple of courses I taught at the University of Strathclyde. They were a bit rough, but thanks to the students that responded to them with some degree of enthusiasm, Fran not least. I developed some of the material on urbanity and cities during four years of teaching at the University of Southampton. Hopefully these were a bit less rough, and now, at least, there is a recommended text!

At Arnold, my thanks go to Luciana O'Flaherty and Liz Gooster who were very encouraging at appropriate points, and to Abigail Woodman for her efficiency in finishing the process off.

The Department of Geography, University of Southampton, was an exemplary place to work, generous in their research support and the time allowed to write projects such as these. Adrian Smith, Alan Latham, and Derek McCormack have commented at various stages on parts of the text, for which I am very grateful. In the wider geography community, James Sidaway and Toni Luna spurred me on by pointing to a need for a book such as this, and I hope it doesn't disappoint. Thanks to Guy Baeten for supplying me with material on Brussels.

Photos are either by me, or drawn from the Audiovisual Library of the European Union, and are copyright of the European Communities. The Cartography Unit of the University of Southampton are responsible for the maps, except for two which are, again, drawn from EU sources, so many thanks to Tim Aspden, Bob Smith, and Lyn Ertl, and to Andy Vowles for his help with the photos.

There is a small amount of material in the book which I have drawn from previously published papers: 'McGuggenisation: globalisation and national identity in the Basque country', *Political Geography* (2000) 19. 473–494; 'Embodying a Europe of the cities: the geographies of mayoral leadership', *Area* (2001), 33, 353–359; 'Rome, global city? Church, state and the Jubilee 2000', *Political Geography* (2003), 535–556.

On a personal level, thanks Mum, as per usual, for your constant love and support. Kimo came on the scene relatively late in the book's production, which is perhaps just as well, and has been a source of immense entertainment. Writing this book was a messy affair, and I have to thank my flatmates in GG for rarely complaining about: having a mouldering old Apple permanently stationed in the corner of our lounge with corn flake crumbs in the keyboard; the piles of photocopies that ornamented the table; and the jumble of books that sprouted as quickly as the banana plant died. So, Brown and World-Weary I and II, thank you.

Acknowledgements

Finally, I want to dedicate this book to the memory of my father, who thought that the UK should join the European Community so that we'd get cheaper wine. We have. A lot of it is Australian, but we spent some great days in Trieste, Venice, Barcelona, and Mallorca being, I suppose, European.

Introduction

The study of the European Union has become one of the major challenges of contemporary social science. At the beginning of 2003, the EU consisted of 378.56 million people, encompassing 15 nation-states, hundreds of cities, multiple languages and deeply-held identities. By the end of the decade, this will have swollen by over 100 million, with the accession of ten new members, most of which are from the post-communist states. Almost every discipline has had its say on this ever-expanding map. Some of these accounts interrogate the meaning of Europe by working through its complexities. Some others take a fixed meaning, 'containerising' it. Yet whatever the approach, in any library or bookshop you will find stacks and piles of books with Europe in the title. My focus here is on the territory of the European Union, a territory that has expanded vastly in recent years, and how we might conceptualise, frame, or *imagine* it in a critical manner. As such, the book is part political geography, part a meditation on the cultural significance of mobility for European integration, and part an attempt to think through the various expressions of European popular culture. It is *not* a systematic or comprehensive guide to Europe, and should be read as a route to further sources rather than as a coherent theoretical model.

As such, the title of this book requires some explanation. To call anything 'new' is to immediately invite the ageing process. In 1968, Anthony Sampson's fine portrait, *The New Europeans*, captured the spirit of an age where the continent's 'novelty' was its sudden emergence from the ruins of the Second World War:

> Driving through Europe in the course of writing this book, twenty years after my first glimpse of the continent, I found the transformation still scarcely believable. I was travelling with my wife and baby daughter, staying in flats in suburbs, surrounded by the sounds of cars and lawnmowers; the cosy society seemed to have no link with the wildness of the early post-war period … All over western Europe a shiny and uniform superstructure has grown up, of supermarkets, skyscrapers, television towers, motorways … Talking to the politicians, tycoons or trade unionists, I found everyone seemed preoccupied by material growth … The continent was cut off from the desperate troubles of the world outside – the war in Vietnam, the racial crises in Africa, or the poverty of India. With the retreat from the empires, and the settling of overseas wars, Europe had turned in on itself, content with its new self-sufficiency
>
> (Sampson 1968: viii–ix)

This portrait of a Europe basking in the affluence of the economic 'miracles' of the 1960s is instructive for several reasons, particularly from the standpoint of a rather less content state of mind. For one thing, Europe then was the Common Market of a mere six states: the so-called Benelux (Belgium, Netherlands, Luxembourg) states, France, West Germany, and Italy. Spain, Portugal, and Greece still laboured under dictatorships, the Scandinavian States followed their own models of self-sufficient growth, and the UK and Ireland looked on from the Atlantic margins with a mix of disdain and envy. By the end of the century, almost all of these countries were in the club, and the countries largely absent from Sampson's odyssey were queuing with determination to become part of the same dream of affluence. The obsession is still with material growth, but the optimism of unbridled expansion is now very much challenged, both because of environmental costs, and by the very simple evidence that the free market creates huge disparities between and within Europe's regions. And, even before 9/11 and the election of George W. Bush as US president, it was clear that the expansion of the continent would 'entangle' Europe's institutions and societies in the post-colonial and post-communist world outside.

My intention in this book is not to simply revisit the 'fact mountain' that study of Europe can entail. Instead, I aim to work between the varied perspectives and literatures provided from a number of disciplines – history, politics, sociology, anthropology, geography, cultural studies, and urban planning particularly – to consider ways of inter-preting European 'space'. By this, I mean both territory – the actual, mappable land that we walk upon – and the rather less concrete notion of symbolic space, understood as land- or city-scapes, and the metaphors, stories, and stereotyping that make up the 'pop geographies' of Europe. This is important, because while some authors argue from the perspective that we are cruising more or less smoothly towards a more integrated future, it could be argued that European identity is based upon an understanding of or encounter with 'difference', often drawn from national or regional histories.

Furthermore, this involves taking a view of Europe that is not presupposed by the existence of the European Union (Delanty 1995). There is a Euro-identity which is to do with a single market or Brussels that is of great, and growing, importance. Yet this co-exists, often uneasily, with a whole set of spatial practices, rivalries, and clashing identities within the EU's territories. Here, the likes of Massey (1999), Pile (1999) or Amin and Thrift (2002) have worked through the idea of the urban as a 'spatial formation'. By extension we can approach Europe as a place that is essentially urbanised, in the sense of being connected to infrastructure grids and linked by train and road. Here, a density of people and things, flows and networks that criss-cross the continent's land mass become fundamental agents of socialisation (Jönsson et al. 2000).

Imagined spaces

I grew up in Scotland, in Kirkcaldy, which is a fairly humble coastal town of around 45,000 inhabitants, with a largely obsolete though once very pleasant pedestrianised High Street and an orbital economy based upon Sainsbury's, PetsMart and Asda. It is on the Forth river estuary, which is the UK's largest port area after London in terms of tonnage handled. When I found this out I was surprised, and a little embarassed, for although I used to register the hulking grey shapes of bulk carriers out on the water from some of the town's higher vantage points, I had no idea of *my* town's relative importance in the UK space economy. This got me thinking about other ways in which I neglected to 'spatially' imagine Kirkcaldy, and wondered if it was really my fault, because at a time when we think of the world as being increasingly integrated, and of the UK as being increasingly Europeanised, this town has probably been less and less so. I knew that Adam Smith was our most famous son, though didn't register how his ideas had travelled to shape those of Karl Marx. I knew the town had old, strong trading links with Flanders, but grew up thinking south, not east, towards London and England, or west across the Atlantic to America. I knew that the town had an important role in the mining of coal and the production of, particularly, linoleum, but didn't stop to think about how this might be exported, as if it rolled straight out of the factory or out of the mine and only on to the floors and the grates of Kirkcaldy's grey stone terraced houses. So I didn't connect, which is a shame, as I do now make these links, yet I no longer live there. I can begin to see my town with a 'geographical imagination' (Allen *et al.* 1998; Gregory 1994), perhaps more easily now I have left.

Such imagination can be aided by making connections between three forms of geographical knowledge. First, there are the visible *landscapes* (buildings, public spaces, industrial sites and consumption spaces, infrastructures and the rural) and different ideas of *territorial shape* (Europe, the nation, the region, the city) that allow us to make sense of space. As I note below, these involve the consideration of 'imagined spaces', how territories are framed popularly, and how these knowledges are formed or guided through the state. Second, there is the issue of *mobility* and movement, whether this be of people or things, and the technologies that allow this, and as a corollary the barriers and obstacles to such mobility. Third, there are the various essences, pop geographies, and other forms of *representation* (travelogues, media place myths, political visions of mayors and prime ministers and regional presidents). As a means of highlighting these connections, each chapter of the book has a box that reflects upon a key space within European territory, and which seeks to localise some of the broader trends discussed. I have identified Brussels (as a potential capital of the New Europe), Euro Disney (as an Americanised threat to national identity), the Bilbao Guggenheim (as an example of how regional movements can position themselves within global contexts), Barcelona's La Rambla and the Vatican in Rome (both central to the importance of a Europe of the Cities), the European railway station (as a space of movement and connection), the Berlin Wall (the epitome of the barrier to free movement suffered by many who desired the economic and cultural opportunities offered under the EU's banner), and finally, the Paris suburbs, location of Maspero's (1994) journey through the hidden landscapes of French society.

So, it is seems clear to me that as a concept, Europe cannot be understood by studying its territory or institutions alone, and without paying attention to works of literature, to political analysis, to planning and architecture, and, of course, to political, social and cultural history. In the writing of this book, a number of works provided inspiration. To name but a few: Drakulić (1987, 1996), Enzensberger (1989), Maspero (1994), Moretti (1999), Robb (1998), Garton Ash (2000), Sampson (1968), Bradbury (1996), Fraser (1998), Ladd (1997), Pascoe (2001), Pred (1995), or Sante (1998) all burst with imagination and ideas, and their influence should be felt in many of the following chapters.

Mobilities and urbanities

> Speculation on the meanings of contemporary Europe have been part of the problem of how to deal with the (rhetoric of) global flows and cultural closure at various spatial scales. These themes have become popular along with Castells's ideas . . . about the rise of the space of flows that will challenge the space of places. In the fields of critical geopolitics and international relations, some scholars have become increasingly interested in this 'fast geography' that challenges traditional forms of territoriality. Most working in continental Europe, however, seem to understand the transformation of the world through a 'slower geography' where territoriality and spatially bounded loyalties persistently shape the politics and daily lives of ordinary people, which nevertheless take place in a globalizing world.
>
> (Paasi 2001: 18)

A second theme of this book is that to understand the New Europe requires us to understand its dual condition of being highly urbanised and highly mobile. By this, I mean – of course – that Europe can be viewed through its cities. It also means that it can be understood through everyday urban lives. But it also means that even rural Europe is 'perforated' by the urban, whether in its infrastructures, its media images, its floods of tourists *escaping* urban life (Amin and Thrift 2002: 1). While I consider a number of the iconic spaces and places of European life – St Peter's basilica in Rome, La Rambla in Barcelona, the Bilbao Guggenheim, for example – I also want to emphasise that there is a powerful myth surrounding notions of a 'Europe of the Cities'. For every Venice there is a Porto Marghera, for every London there is a Watford, for every Barcelona there is a Benidorm. And my concern is not to ridicule these places – quite the opposite, in fact. It is in these urban spaces that much of the movement of Europe is taking place, where identities are vivid, where churches may be fuller or poverty may be greater. It is these spaces – on everyone's individual map, but off the collective imagination – that should be teased out as well as their more famous, culturally resonant, twin cities.

Essences and stereotypes

Third, it is important to consider that popular knowledge of other places and the 'ways' of their inhabitants are often formed by various forms of cultural encounter, which may be first-hand (through tourism, for example) or through mass media representations. Both may engender a – deliberate? – lack of cultural understanding. In the European case this is magnified due to linguistic barriers, not helped by a long history of nationalised conflict. The German political theorist Hans Magnus Enzensberger, in his travels through Europe, is irate at such comedy.

Typically Italian. Which is it then? The opera or the Mafia? A *cappuccino* or bribery? ...Whenever anyone says that something or other is 'typically Italian,' I want to jump up with impatience, overturn my chair, and run out of the room. Could anything be more barren than the study of 'national psychology,' that moldy garbage heap of stereotypes, prejudices, and accepted ideas? ...And yet it is impossible to dislodge these traditional garden gnomes with their naively painted faces: the taciturn Scandinavian, blonder than straw; the obstinate German, beer stein in hand; the red-faced, garrulous Irishman, always smelling of whiskey; and, of course, the Italian with his mustache, forever sensual but regrettably unreliable, brilliant but lazy, passionate but scheming.

(Enzenberger 1989: 76)

Of course, Enzensberger has a point, even if his constructions may not be fully shared by all of us, who have our own, updated stereotypes of national (and regional) characters. It is important to remember that societies are full of stereotypes, and that the press, politicians, and romantic-minded travel writers, or advertising agencies, or Hollywood producers are happy to permeate the individual consciousness with calculated fictions designed to win votes, sell books, brand products, or get us into the cinema. In many ways, this forms part of Billig's (1995) idea of 'banal nationalism' where our geographical understanding are formed through an array of mundane practices, habits, jokes and stories.

Position

Finally, there is an enormous problem facing any academic work that discusses Europe: that of standpoint. In such a huge and diverse continent, how can one aim to either be comprehensive in scope, or devoid of national (or other) prejudice? Stenning (2000) reflects on this in a recent review of three textbooks on the 'New Europe' (Budge and Newton et al. 1997; Pinder 1998; Unwin 1998) arguing that some of these works (particularly the former two) only half-heartedly address issues of *position*. 'Until recently' she continues, 'it was easy to justify writing textbooks about just one half of a divided continent ... However, it is no longer possible to write of "Europe" without considering *all* its many parts ... and this historical fact poses, as we can see, huge challenges for editors and authors alike' (Stenning 2000: 111, emphasis added). For some textbooks, she continues, 'the "new Europe" is in fact the old, Western Europe with east-central Europe tacked on almost haphazardly' (p. 111). This is a very real problem. The idea of comprehensively 'covering' Europe becomes an impossibility, and requires tough decisions on how to represent the considerable social and spatial diversity of the continent. This book contains very little material on East-Central Europe, ignores Scandinavia bar a brief excursion to Stockholm now and then, barely mentions Austria, and isn't too hot on the Mediterranean either. Yet my hope is that through a focus on shape and mobility, an understanding might be gained of how Europe might be described avoiding such dominating frameworks.

The whole nature of framing Europe as an object of study is elaborated at length by Norman Davies (1996) in his introduction to *Europe: A History*. Davies recalls attempts to write a 'common European history textbook' in the 1980s, with funding provided by the European Commission:

> Labelled 'An Adventure in Understanding', it was planned in three stages: a 500-page survey of European history, a 10-part television series, and a school textbook to be published simultaneously in all eight languages of the EC .... The text [of the latter] had been established by collective decision. A French account of the 'Barbarian Invasions' was changed to 'the Germanic invasions'. A Spanish description of Sir Francis Drake as a 'pirate' was overruled. A picture of General de Gaulle among the portraits on the cover was replaced by one of Queen Victoria.
>
> (Davies 1996: 43–4)

This book does not aim to follow such an approach. Its biases will become obvious, considering that I am a London-dwelling Scot immersed in British popular culture, but with a research career that has involved stints in Spain and Italy, or more properly, Barcelona and Rome. My aim is rather to identify frameworks for understanding how we might represent, think about, or even 'map' Europe as a mobile space full of richly evoked places.

So, we can think about how our imaginings, our means of understanding Europe and its component parts, are something to be explored and unpacked, rather than simply derided. The idea of the 'other', the way we think of 'foreigners', is based on constructing an idea of homelands, be they Europe, the nation, the region, or the city.

Structure and contents

The book is arranged around seven chapters. I begin by exploring the idea of the 'Europeanisation' of Europe, which seeks to position Europe as an ideology, as an actual way of thinking about culture and territory, rather than a mere adjective to describe a grouping of neighbouring nation-states. Here, I outline the milestones in the development of the European Union, before exploring the attempt to develop Europe as a 'cultural project' (Shore 2000). The chapter critically examines policies and areas where European integration might proceed or be stifled, including the adoption of the Euro (as a single currency), the role of Brussels both as a potential 'capital' and as a bureaucratic centre, and the range of policies and artefacts used to 'invent' Europe as a political and cultural community. However, a key aspect of this is the differentiation of this community from 'others', such as Islam, the United States, and the anxieties involved in its enlargement eastwards.

In many ways, the formation of the EU is similar to the foundation of nation-states. Governments need to have their jurisdiction recognised both by their residents, and by other governments. The pace of European integration has been seen to hasten the decline of national sovereignty, to the extent that some critics point to the 'death of the nation-state'. Yet the story is more complex than that. In Chapter 2, I review the uneven response of various nations to this challenge. It can be argued that some countries –

Ireland and Spain, for example – are ardent Euro-enthusiasts, seeing it as a means to modernise their economies and cultures. Other governments have been more cautious, but have sought to rebrand themselves within a competing 'territorial market', both as a means of seeing off internal enemies, but also to improve their 'brand recognition' more generally. Each of these responses involves the mass recognition of nationhood, whether in the sports people follow, the cars they drive, or the iconic landscapes and cityscapes of their nation. I argue that the development of closer integration will involve a repositioning of national cultures and identities, but will not undermine the nation-state necessarily.

In this context, it is perhaps wise to avoid grandiose statements of the future of states and identities in Europe. For example, James Anderson (1996: 135) has argued that 'the presentation of a simplistic "choice" between just two alternatives – life or death – obscures the possibility that something else is happening: a qualitative reshaping of states and nations, territoriality and sovereignty, which is not captured by notions of death or decline'. Here, there is a whole list of agents jostling with nation-states for power, or for the loyalties of their subjects and citizens, from transnational corporations to city and regional governments, from religious and linguistic networks to the bureaucracies and legalities of institutions as diverse as the EU, NATO or United Nations. According to Anderson (1996: 135), this suggests that 'political spaces [are] becoming at least as complex as those of medieval Europe'. This idea is often summarised in the notion of a 'Europe of the Regions' nestling within the embrace of a federal EU, an allusion to the situation of the Middle Ages where religious, territorial, and economic powers meshed and intermingled. So rather than one, all-powerful state controlling a single territory, theorists such as Bull (1977) foresee a 'neo-medieval' model where the likes of Scotland or Catalonia will gain additional powers and freedoms to dictate cultural norms, within a more general 'world' or European system of government. Yet the idea of a regional Europe is often loosely used, and in Chapter 3 I critique this notion through a discussion of several contemporary case studies.

Chapter 4, 'Europe of the Cities', might appear to continue down a 'scalar staircase' (Europe, nation-state, region, city) but the intention of this chapter is to position the city at the centre of our understanding of European life. Furthermore, it is in cities that we see the complex transnational connections that disrupt notions of stable, fixed territories. Rome, particularly, continues to exert a powerful influence over millions of Catholics worldwide; national capitals as diverse as Madrid, Amsterdam, and Lisbon continue to demonstrate the complexity of migration flows and post-colonial relationships around the globe. Furthermore, the disruptions brought by economic restructuring and globalisation are increasingly changing the shape and cultures of these cities, leading some to react with anger or nostalgia over the loss of familiar landscapes and ways of life. Above all, the European city is poised between the sense of possessing a fixed 'essence', with stereotyped inhabitants, and cities and networks, linked by time and space to other cities both in Europe and around the world, perhaps more than to their own hinterlands.

Here, civic representatives such as mayors, and civic institutions such as football clubs, have the power to project the image and *name* of the city.

The idea of the network leads into Chapter 5, which stresses that the EU owes its legitimacy to mobility and (preferably fast) movement. The most vibrant European consciousness is gained by the diverse groups that – either regularly or temporarily – flood across Europe's lands, as tourists, students, economic migrants, terrorists and criminals, or in pursuit of business or professional goals. To enable this, the continent is stitched together by a network of roads, ports, railways, and airways, in which the increasing demand for speed is fostering a significant amount of infrastructural investment. By contrast, Chapter 6 is about barriers to movement. The metaphor of Europe as a 'fortress' competes with analogies of European integration as a 'bridging' process (Pollard and Sidaway 2002; Sidaway 2001). New, post-national borderlands emerge that both foster transcultural understandings but also, perhaps, increase the risk of encounter with the 'other'. However, as Walters (2002a, 2002b) and others have argued, it is not clear that Europe's borders are now the fixed lines on the map that we are familiar with. Borders may now be far less tangible, expressed in technologies, artefacts (the possession of a passport or visa), than those 'socially constructed' by schoolbooks and politicians.

Such an emphasis on mobility draws the book into its final argument, which is intended to act as a springboard for further research. The basic argument here is that the EU, for all its lofty proclamations about co-existence and citizenship, will succeed or fail in its ability to guarantee its citizens the rising standard of living experienced by post-war generations. In particular, the enjoyment of a 'Western' choice of consumer goods was a prevalent interpretation of the decline of communism in East-Central Europe. And the continuing expansion or enlargement of the EU eastwards will bring this dream (or right?) to an additional 100 million people by the end of the decade. This will result in a new European landscape unlikely to be on any tourist's schedule, one that can't be caught in a postcard, and one that has a very different image than a Van Gogh or Cezanne. Yet it is here, in the urban hinterlands of Bratislava, Campania, or the Ruhr, that we find the vibrant landscapes of the New Europe: huge warehouses of mass produced German brie, factory farms providing the continent's battery chickens, eggs, fish fingers and bacon, massive container ports such as Rotterdam Mainport, edge city office developments, huge airport cities, bulk retail outlets, miles of roadscape dedicated to the continent's unsustainable future lined with car showrooms, lay-bys, distributor roads leading to factories, scrapyards, and servicing centres. Understanding this economy and geography, understanding the social lives of commuters, security guards, lorry drivers, and working out how best to map and represent this Europe, is the challenge posed by Chapter 7, a suitable endpoint to a Europe caught between fixity and mobility.

1
The Europeanisation of Europe

The fundamental dilemma for the EU lies in the fact that the 'European public', or *demos*, barely exists as a recognisable category, and hardly at all as a subjective or *self-recognising* body – except perhaps among a small coterie of European politicians, administrators and businesspeople.

(Shore 2000:19).

If we accept that spaces are imagined, socially constructed, and endlessly represented and consumed, then we cannot assume that 'Europe' has a pre-existent identity. In the run-up to the 2003 Iraq war, the US Defence Secretary, Donald Rumsfeld, characterised France and Germany as 'old Europe' due to their refusal to support the US invasion. Such a remark provoked a diplomatic storm, but – above all – demonstrated the power of words in shaping political practice. Similarly, in its attempts to replace the nation-state, the 'European Union' is by no means a neutral title, as even in this naming process, politicians and civil servants are trying to give social meaning to a particular political territory. They are doing so in a way that is familiar to any nation-state over the last 100 or so years – through a process of active *building* or *construction*. Just as most social scientists accept Benedict Anderson's (1995) account of national elites actively forging an 'imagined community' where citizens are created through education or voting, so Europe is being built day by day. Effectively, Europe is being *Europeanised* – and this process is being resisted just as it is being pushed forward.

So, this chapter proceeds from the understanding that 'Europe' is an imagined community, that it has been 'captured' by the European Union as a means of enhancing the influence of its most dominant nation-states over world affairs, and that Europe is a project, in need of construction, a community in need of imagination. Here, Cris Shore's (2000) fascinating anthropology of the European Commission, *Building Europe,* has provided many missing clues as to how this has been happening. But in addition, political scientists such as Peter Van Ham (2001), or journalists like Anthony Sampson (1968), have in diverse and fascinating ways tried to map out exactly where this has happened and where it is succeeding and failing.

The chapter has four parts. It begins by briefly charting the process of European integration in post-war Europe to the most recent manifestations of unity, the launch of the Euro. I then move on to discuss how this has been actively orchestrated, both by self-consciously 'European' civil servants, as well as by national politicians driven by diverse motives. How can the Union's 378 million inhabitants – the majority of whom identify most closely with the nation-state – be urged to locate themselves within a new European superstate? Third, I discuss the emergence of a 'geography' of European governance based around Brussels, and consider whether the city can function as a capital of Europe. Fourth, I consider how this Europeanisation process has had to contend with three major external forces, America, Islam, and the post-communist East and Central Europe, which are often presented as the 'others' against whom Europeans might identify themselves collectively (Van Ham 2001) in an uncertain post-Cold War era.

Europe as a political project

The evolution of the European integration process after the Second World War is a well-known story, and I am not going to repeat it here. Almost any textbook will provide a discussion (eg. Guerrina 2002; Nugent 2002). However, it is important to consider briefly some of the stages of this process, particularly from a geographical stance. Two concepts are central here: the idea of mappable *territory,* and the idea of *sovereignty* (crudely, the ability of governments to have absolute power in regulating their own economies and societies, and to resist external forces, be these cultural flows, economic policies or military intervention).

Some of the key milestones in the integration process are as follows, and my discussion of them is deliberately provocative and even tendentious.

1945 The end of the Second World War leaves the continent shattered. The idea that apparently 'civilised' nations could submit their citizens to mass death jostles with the equally convincing lessons drawn from the rise of the Nazis: that democratic rights cannot be taken for granted; and that large corporations (which helped finance Hitler's government) generally act to maximise their profits, and are likely to undermine democratic rights where their interests are threatened. Similarly, they may support democratic states where these help profit margins. The whole evolution of the ideal of an integrated, unified Europe grows within this tension – corporations welcome the idea of a single market, a unified Europe seeks democratic mandate. Can the twain meet?

1957 The Treaty of Rome is signed, article 2 of which calls for 'a harmonious development of economic activities, a continuous and balanced expansion, an increase in stability, an accelerated raising of the standard of living and closer relations between the States belonging to it' (extracted in Leonard and Leonard 2001: 26). At this stage, there are only six members, Italy, West Germany, France and the three 'Benelux' countries, Belgium, Netherlands, and Luxembourg.

1961 The Cold War, which has rumbled chillingly by since the post-war division of Europe at Potsdam and Teheran, sees its most fundamental icon built: the Berlin Wall.

1973 Britain, Ireland, and Denmark join. Britain had in the 1950s stayed out of the original group of six, and immediately began to fear a major policy error. In 1963, its attempts to join were barred by French president Charles de Gaulle's veto, which was both tactical on his part (retaining French diplomatic strength) and culturally significant, helping to reinforce Anglo-French differences, and reinforcing the notion of the 'island nation' and the 'continent' that resurfaced on the building of the Channel Tunnel.

1981–86 The Southern flank. The accession of Greece (1981), then Spain and Portugal (1986), all of which had laboured under dictatorship until the mid-1970s, provided a flow of Euro-enthusiast politicians and civil servants who knew at first hand what democratic rights meant. Simultaneously, the Community now had borders with countries that challenged the idea of a 'European' as being 'white and Christian'. Territorially, the land size of the EC was hugely increased, and the economic burden of integrating these weaker economies was raised.

1989 The fall of the Berlin Wall has a fundamental bearing on the future evolution of the EU. The reunification of Germany in 1990 gives that country a hugely expanded role in the development of the continent, both by its new size and its geographical location. Mikhail Gorbachev's speech to the Council of Europe in July of that year had cannily opened up expectations of an end to the Iron Curtain: 'as far as the economic content of the common European home is concerned, we regard as a realistic prospect – though not a close one – the emergence of a vast economic space from the Atlantic to the Urals where eastern and western parts would be strongly interlocked' (extracted in Leonard and Leonard 2001: 102).

1992 With the signing of the Maastricht Treaty, the European Community becomes the European Union, but more importantly, the basis is laid for a whole new transfer of competencies from the nation state to the EU (channelled through the nationally-composed Council of Ministers). Areas from culture to defence are included, but most important is the kick-starting of a process of European Monetary Union (EMU), which aims to consolidate the harmonisation measures set out in the 1986 Single European Act. Here, the long-cherished idea of a single European market and economic space is brought into being (Euroland), which alters the whole logic of seeing the continent as a 'family of nations'.

A significant, if hardly central, innovation is the establishment of the Committee of the Regions (COR), which has representatives of regional and city councils across Europe. The body has no decision making power, only an advisory role, but to its advocates may represent the first stage towards a 'Europe of the Regions' (see Chapter 3), a possible model of Euro-federalism.

1999 The Commission's mode of operation is rocked by corruption allegations, leading to resignation of Commissioners *en masse*. In the process, they fulfil almost every ill-informed prejudice of the most rabid English Eurosceptic. The European Parliament's report into the lax practices contained the phrase, 'it is becoming difficult to find anybody who has even the slightest sense of responsibility' (Leonard and Leonard 2001: 118). President Jacques Santer and his 19 commissioners resigned, and Romano Prodi became the new president, whose term in office would include the launch of the most significant step towards an integrated Europe hitherto seen.

Figure 1 _Jacques Chirac with Romano Prodi. © Audiovisual Library European Commission._

2002 The Euro comes into circulation on the 1st of January (E-day), perhaps the most significant step towards European integration yet made. As the culmination of the EMU project which was set out in the Maastricht Treaty, the intention was to harmonise macroeconomic policy among the member states, thus flattening out the idea of a single economic space. With economic sovereignty being located with the European Central Bank in Frankfurt, one of the most important tools of national governments – the ability to vary its interest rates – is removed. There are only a couple of clouds. First, a number of member states opt out of membership, (UK, Greece (which later joined), Denmark, and Sweden), thus limiting the scope of the project. Second, it is unclear how the currency will operate in practice, with some feeling that it will lead to a de facto German dominance of economic policy. Yet, as I discuss below, perhaps the biggest impact of the Euro will be in its formation of a European consciousness, given the everyday importance of currencies and money.

2003 The Nice Treaty comes into force. This Treaty – which paves the way for the enlargement of the EU from 15 to 25 over the coming years – was significantly rejected by the Irish electorate in a referendum in June 2001. Although subsequently passed in a second referendum in 2002 (on a 48 per cent turn-out), the earlier rejection symbolised the response of a Euro-enthusiast public that had benefited economically from EU membership, and who feared a dilution of that success when new central and eastern European members were admitted. Thus the Irish referenda symbolise the problematic nature of belonging and European identity, tensions which may be increased in coming years.

Figure 2 *Map of EU member states (including applicant states with projected entry dates).*

So, these milestones to integration have seen the whole idea of Europe being transformed, as actual political and economic integration has proceeded apace. Yet my intention in this chapter is to explore how this expansion through the EU (which has arguably 'stolen' the idea of Europe) has seen the construction of a particular European space and identity.

The invention of Europe

So, given that the idea of 'Europe' as a political and economic project is largely elite driven, by both a Brussels-based civil service and national government ministers, it is not entirely clear how the gap between the 'people' and the project can be bridged. This – often known as the 'democratic deficit' – remains a source of worry for many European leaders, though usually only when referenda on ratification of treaties register rejection or only lukewarm support, as with the case for the Danish and French votes on Maastricht, and the Irish rejection of Nice. Part of the problem lies in the question of European identity. The history of the term is long and tortuous (see Norman Davies's *Europe: A History* (1996) or Heffernan's *The Meaning of Europe* (1998) for discussions of the emergence of the Europe 'label'). And what do we mean when we speak of the 'New Europe'? Here, I break the discussion into three parts: first, the attempt by the European Commission (which has a complex interaction with elected national and European politicians) to symbolically 'build' a Europe that ordinary people can identify with; second, the intensification of this process through the introduction of a single European currency; third, the parallel existence of what we could call a European popular culture, one that develops out of the clash of national cultures in various spaces of encounter across the continent.

Symbolic construction: Europe of flags and anthems

So how has this New Europe been constructed symbolically? Adopting an anthropological approach, Shore (2000) identifies a group that he calls the 'agents of European consciousness', by which he means 'those actors, actions, artefacts, bodies, institutions, policies and representations which, singularly or collectively, help to engender awareness and promote acceptance of the "European idea"' (p. 26). By this, Shore means both human and material objects and actors – the feel of a Euro coin in your pocket, or the sound of an Erasmus student's flawed English, or, indeed, the intangible and foggy policies that emanate from 'Brussels' and which alter aspects of everyday life for good. This, he suggests, is a 'banal' imagining (a term borrowed from Michael Billig (1995) that we will meet again in the next chapter). Yet he does *not* appear to mean the existing elements of 'European' popular culture in which we might place footballers, Spanish lager, the summer romance between a Swedish girl and a Greek waiter, a Fiat car, or any other number of travelling – mobile – objects and people.

To be clear, Shore is arguing that the European Union has a whole set of agents that are part and parcel of a *Europeanisation process* and which are dedicated to the building of a supranational state – in other words, a form of governance that re-orders (but is unlikely to replace, at least in the foreseeable future) the scales of government that we take for granted at the moment (see also Bellier and Wilson (2000)). Yet the lack of any permanent or stable sense of identity or belonging has been addressed by the European Commission through a variety of initiatives, most of which are based on 'invented traditions' (cf. Hobsbawm and Ranger 1992), a practice borrowed from the nation-state of the nineteenth century. What might these include?

- *Mass educational initiatives* such as the ERASMUS and SOCRATES exchange programmes may help to foment a sense of trans-European friendship and sensitivity to other cultures among students, as well as encouraging transnational research collaboration (Murphy-Lejeune 2002).

- *The Structural Funds*, such as the European Regional Development Fund (which part aids new infrastructure developments such as motorways), Interreg (which supports co-operation between border areas of nation-states) or the Common Agricultural Policy (the CAP, which is perhaps the most inflammatory of the European funds) have raised the EU's profile enormously, as they have offered 'free money' to those able to find projects suitable for funding. They have also provoked international rivalries over the relative share of contributions, and prompted hostility to the funding shift that is already taking place as the EU prepares for enlargement.

- The *European Flag* – blue with a gold star representing each of its members – is now a prevalent part of the visual landscape in member states, whether on car bumper stickers, flying outside public buildings, or ornamenting the notice boards accompanying development work financed from EU Structural Funds.

- The standard format purple European *passport* was introduced to mild grumblings from some, but without major opposition. As I describe in Chapter 6, however, passports have material functions that allow or block mobility within the new European 'borderless' space.

- The introduction of *Eurostatistics* through *Eurobarometer* (an opinion poll on various issues surrounding integration) or *Eurostat* (which effectively acts like a national statistics office for the EU) is an essential first step in shaping Europe: these are

> not only powerful political instruments for creating a knowable, quantifiable and hence more tangible and governable 'European population' and 'European space': rather, they are also powerful moulders of consciousness that furnish the meta-classifications within which identities and subjectivities are formed.
>
> (Shore 2000: 31)

By this, Shore simply means that – as with most nation-states a century or so before them – the creation of categories of measurement are the first step in formulating policy, and have always been a significant part of the state-building process (see also Walters 2002b).

- The creation of a transnational, or post-national, *'European map'*. Here, the use of spatial metaphors such as the 'blue banana' (a corridor of intense economic growth that can be drawn from London through Brussels and Paris, post-industrial, high-technology areas of Germany, to Milan and Lombardy) are, according to Jensen and Richardson (2003) and Kunzmann (1996) important 'visioning' mechanisms where policy-makers 'think' European. For Sparke (2000: 203), 'the old geopolitical imaginary is completely erased and in its stead comes a new kind of geoeconomic cartography, mapping not armies but spaces of economic growth, not trenches of conflict but zones of cooperation and commonality'.

Yet these measures are necessarily limited. On the 1st of January, 2002, we saw the biggest move towards a technology that fused both symbol and everyday life – the entry into circulation of the single European currency, the Euro.

Europe as money: the Euro

It is often argued that a country's bank-notes provide one of the most common means by which a national consciousness is embedded through the daily use of currency in mundane transactions (Pointon 1998). Money is a powerful source of everyday identity – think of the numerous slang words that we use in its place, in any of the major languages. As Shore remarks:

> Coins and banknotes have traditionally functioned to define the boundaries of kingdoms, empires and nations. The very first coins bore the profile of emperors and kings and symbols of state power, much as they do today. With their images of national figures, monuments and inventions, banknotes and coins are powerful icons and instruments of government that help to render abstract ideas of state and nationhood a daily political reality.
>
> (Shore 2000: 91)

From buying bread or petrol, to discussing property prices, to the requests of the street beggar, to the opening of the pay or dole cheque, monetary conversations are a central part of everyday culture. And this is often tied to the performance of the national currency. In Central and Eastern Europe, the post-communist states have overwhelmingly chosen to look back to the past in choosing symbols for their banknotes, which often feature (male) historical figures from the nineteenth century (Unwin and Hewitt 2001). The UK Conservative party entered the 2001 general election with its leader brandishing pound coins and warning of their imminent disappearance, should the Labour government be re-elected.

For Europeanists seeking a tool to forge a unified identity among diverse cultures, the answer was not to be found in a common language like Esperanto, but in a single currency. And so the design of the Euro – with nationally-specific designs on one side and a common European symbolism on the other – will, arguably, provide a gradual means of socialisation for all those who use it (Shore 2000: 111–118; Van Ham 2001: 76–77). As Pollard and Sidaway (2002) have suggested, it is no accident that the appearance of bridges on the rear of Euro-notes act as metaphors of 'connection, communication and openness' (p. 7). Yet they go on to argue that the introduction of the Euro masks the fact that the integration project will have enormous implications on the peoples of Europe in a deeply uneven way – the macroeconomic policy of European monetary union may, for example, exacerbate the relative poverty of 'lagging regions' on the periphery, thus eroding the legitimacy of the integration project.

Yet it remains true that despite the functioning of the Euro as a casual mode of European consciousness-building, the decisions which drive the Euro's value and performance are being taken at a level ever more remote from European citizens. Dyson (1994: 5, in Shore 2000: 91) has noted that 'banks have a symbolic as well as technical importance, often expressed in their imposing facades – for instance, those of the Bank of England and the Banque de France'. Jacobs (1994, 1996) has described how plans to redevelop a site opposite the Bank of England in London revealed a deep-seated fear of 'foreign' architectural influence and a defensive reaction to the decline of British imperial might. In this context, the creation of a European Central Bank in Frankfurt, along with the Berlaymont building the epitome of faceless, technocratic Europe, embodied British fears about control by German bankers (Morrison 2001: 6).

So, one of the major obstacles to generating Euro-enthusiasm in countries such as Britain is overcoming the aura of bureaucratic dictatorship that many – rightly or wrongly – fear. Given this problem – that European identity has until now been largely a techno-cratic and economic programme – it is not clear that a functional European identity can be easily forged. As Van Ham concludes:

> It is therefore doubtful that the introduction of a European anthem, a European Flag, European monuments for the 'glorious dead', European ceremonies, universities and museums for 'European heroes' or Europe's 'Founding Fathers', will generate the feeling of historicity, of common roots and belonging. Even in an era where the image has become the principal method of collective appeal and public address, the notion of Europe has so many competitors that it will be difficult to capture a market niche in the collective consciousness of European society.
>
> (Van Ham 2001: 77)

Yet it can be argued that a European consciousness does already exist, if we are prepared to look at the mundane and even bizarre international cultural minglings that have characterised post-war European economic expansion and cultural integration.

Europe as popular culture

Despite the flag, the anthems, and even the structural funds, the most dynamic forms of European cultural identity are not emanating from Brussels, but arise rather more spontaneously. In football, for example, the transfer of European star players from club to club, the cultural intermingling both friendly and hostile offered by major international tournaments (the 1985 Heysel disaster being a fundamental example of the latter), their importance for European broadcasting (driven by corporate advertising and sponsor-ship), has created a clear sense of European belonging (if not unity). Budget air travel and the inter-rail ticket have created an under-researched cultural circuit of tourism. And the fact that images (film) or music cross frontiers with ease has given us the curiosity of the Eurovision song contest, which despite its rather cheesy reputation has proved a long-standing broadcasting success. For Sampson:

> It is quite fitting that the most successful Eurovision link-up should be the preposterous song contest, the annual climax of fantasy and romance. Here at last is an event which the whole continent witnesses collectively. In the 1968 contest, held at the Albert Hall, two hundred million people ... watched singers from seventeen countries ... Most of the songs conformed to their national caricature: all the French-speaking songs were about l'amour, the Italian was about il sole, the Norwegian was about stress. The winner was a

Spanish song with no language barrier, called 'La-la-la', sung by a jolly little girl with huge smiling teeth, who looked like a simple peasant but who turned out to be the daughter of a millionaire impresario.

(Sampson 1968: 305)

Away from Eurovision, a series of songs popularised in the holiday resorts of Southern Europe have gone on to become 'European' hits, where the likes of the Vengaboys, Whigfield, or Los del Río have been imported back into their native countries by holiday-makers.

A European foodscape?

One of the key areas where the EU has sought to intervene is in the area of food and drink consumption. I suggest that here the idea of a European food is based on two things: first, a 'basket' of nationally-stereotyped, mass produced foods, that are *collectively* seen as representative of a European family of nations; second, the creation of an array of lowest-common-denominator foods that, in many ways, are Frankenstein-like (even if not genetically-modified). Here the burger is of course the icon, but we could add the frozen pizza, the cola drink, the 24-hour baguette, and so on. In many ways, this is part and parcel of an 'Americanisation' of Europe that occurred with the post-war penetration of American multinationals into European markets. Yet – as I discuss below – it can also not be dissociated from the 'indigenisation' of such food products in different national contexts.

The European Union (and its predecessor the EC) has become notorious in the UK for its desire to intervene in food production and standards:

Currently maligned policies such as subsidies and set-aside, and the weird landscapes of butter mountains and wine lakes, reflect a desire to regulate farming across Europe, managing surpluses, controlling prices and warding off hunger – as well as shoring up the image of meddlesome Eurocrats who want to mess with everything that's good about British food and farming. Countless urban myths (and some truths, often exaggerated) circulate about EU directives on bananas (giving us the regulation, uniform, *straight* 'Eurobanana') and potato crisp flavourings (no more 'hedgehog flavour' crisps unless there's *real hedgehog* in there), about Mafia money-making through imaginary farm subsidies, and abut mineshafts and landfill sites filled with rotting tomatoes and peaches to keep prices up.

(Bell and Valentine 1997: 113)

But here we have a paradox. The EU is attempting to standardise (for purposes of globally-directed market regulation) the foods that are often taken as nationally-specific, part of the tradition that – invented or not – marks out nations. Yet these national traditions are formed through centuries-old processes of transnational flow.

So, if food – its consumption, particularly – is one of the key markers of national identity in Europe, how do we see the EU fulfilling its dream of inculcating a spirit of belonging and community? In many ways, Sampson's (1968) nightmare vision still prevails:

> But if there is one single new European food, it is – frozen fish-fingers. First invented in 1955, the fingers swept first through Britain, then to the continent, with special success in land-bound countries like Austria, where fresh fish takes three days from the coast. The fingers are the same everywhere, fish sticks in Belgium, bastoncini in Italy, fischstäbchen in Germany.
>
> (Sampson 1968: 222)

Here, a 'Euro-commodification' of mass-produced cans of Guinness, croissants, pizza, and bacon may be one aspect of a generalised globalisation of agriculture and food distribution.

However, I think it is also possible to speak of a 'global' food which corresponds to a different paradigm of European territory – that of an urbanised food distribution and consumption system. Care is needed here, because I think this has two separate – yet linked – meanings. The first is the *corporate* global, the spreading of standardised commodities – be it instant coffee, pizza, and burgers – which are often, but not always, part of a fast food dimension, either in the home (microwave/kettle) or in the street (burger bar or pizza chain). Here, as Fantasia (1995) describes, the French experience of fast food has involved both the French ownership and vending of burgers, but also the adoption of the 'fast food formula' to French products, through the wide spread of the 'viennoiserie', selling traditional staples such as brioche, croissants, and so on (p. 207). Moreover, fast food has 'helped spawn a large institutional food industry (the establishment of canteens and cafeterias along highways, in workplaces, schools and hospitals), and have contributed to the rapid growth of "Agro-Alimentaire", a massive joint agricultural and food-processing industry engaged in the preparation and distribution of frozen and pre-prepared foods' (Fantasia 1995: 205).

The second meaning is the global as a shorthand term for 'immigrant food' - particularly that brought from outside the EU or of US provenance. Thus the kebab, curry, noodles, sushi have been naturalised through a complex process of migratory patterns, ethnic businesses, and their subsequent adoption and diffusion by major food corporations and a web of television food programmes and advertising (aware of the 'changing palate' of the average consumer after gradual exposure to unfamiliar food texture, taste, and appearance). Here, of course, the search is on for the 'authentic' food experience, which generally refers to the ability to buy the kind of food found at street bars in Kuala Lumpur or the downmarket sushi bars of Japanese cities. Again this is often *indigenised* (Appadurai 1990) in a number of ways, either downgraded (the formula curries based on ready-prepared sauces found in many British curry houses) or upgraded (the idea of peasant food being sourced at origin and then cooked for wealthy city dwellers, as in London's *River Café*), or otherwise inflected (sushi arrived in London in two waves, one Japanese migrant, one Californian/Australian-influenced) (Bell and Valentine 1997).

So, in the sense of food and drink being part of a 'performed' identity, what conclusions can we draw? I suggest there are three: first, the nations with strong food cultures are consumed by their neighbours, either through tourism, migrant businesses, but – most pervasively – through the marketing of (especially) 'Italian' foods; second, the existence of a 'global' food – which has two forms, one corporate, one based on non-European immigration – where national citizens are, in fact, consuming a fairly standard product with strong cultural markings of an exotic, possibly far-away place; third, the rise of a European foodscape, where the logic of a single European market has led to a series of regulations and policies (from the Common Agricultural Policy to the straight banana) that have altered the way in which food is consumed and produced under the European flag.

Holiday driving

The idea that European popular culture may be driven by its interaction with other national cultures can be seen in the set of practices and discourses surrounding car use, ranging from attitudes to recognisably national products – Ferrari, BMW, Rolls Royce, Skoda – to the way in which different nationalities actually drive. Holiday driving, in particular, opens up numerous spaces of encounter in this way. Michael (2001), summarising Middleton (1995), highlights the differing driving cultures of Italy, Germany, Spain, France, and Greece:

> The key issue is what makes these stereotypes comparable – the same but different. Minimally, these rages all take place in relation to roads and cars, but the triggering mechanisms are somewhat different. Or rather, in each country different cultural values pertain: the mores, realized in the social conventions of what is to count as an affront and what a commensurate riposte, vary. Thus in Italy it is hesitancy that triggers noisy horn-blowing protest, in Germany it is slowness that triggers headlight-flashing and tailgating, in Spain it is overtaking that prompts re-overtaking and sudden braking, in France it is being English and behaving like the French that necessitates vengeance, in Greece it is inappropriately following the rules that demands histrionic argumentation.
>
> (Michael 2001: 69–70)

Michael's intention is to highlight the fusion of the car as manufactured object with their owner. We might reflect upon the (gendered) national dimensions of the cars that fuse with their 'national' drivers, be it the German 'Opel Man' or the British 'Sierra driver' or 'white van man'. It is this 'sameness with difference' that characterises the European experience: we are tolerantly aware of the habits of our European neighbours, but use that as a marker of how they do things differently.

So, the emergence of a national(ist) car culture, explored in Chapter 2, has been one of the clearest examples of how the apparently homogenising force of the automobile in fact acts as one of the key markers of difference in European society. Despite its enhancement of mobility and mixing, it has been one of the key shapers of the modern nation (state). And while Sampson (1968) notes that from the 1960s, the major European manufacturers were merging within countries, he predicted that 'the bigger the mergers inside countries, the less likely seem mergers across frontiers ... The prospect of a "European General Motors" seems more distant than ever … Europe seems to be settling down quite happily to a car nationalism, which cannot be regarded as very menacing, and ensures a diversity of a kind; alongside the mass producers there will still be room for small companies, like the Swedish Volvo or the Czech Skoda, to join the traffic jam' (Sampson 1968: 133–4). That this prediction has failed the test of time may be one indicator of the way in which the car industry shows the rapid impact of globalisation and integration on national markets.

So, Europe has a vibrant and complex popular culture that impacts upon *national* identities in various ways. However, despite the fact that some European footballers have become household names, or even heroes, chains such as Benetton and Bata are established in high streets from Barcelona to Berlin, and Mediterranean 'delicatessen' foods such as olives and baguette are now staples of many a northern European household diet, this does not necessarily communicate to the level of political consciousness.

> People consume products but their cultural meanings vary according to context. The British preference for Indian food or German cars does not lead inexorably to an identification with India or with Germany. English football fans may worship Eric Cantona or David Ginola but still dislike the French and Italians. The question is, at what point does experience of Europe spill over into 'European consciousness'?
>
> (Shore 2000: 229)

The process of identity formation is a complex one, then, and the diversity of European iconic 'bundles' may be easier to consume where no political responsibility exists. Where else can we look for sources of democratic identity? Many citizens have strong mental images and emotional responses to their capital cities. But does this apply to Brussels?

Box 1 Brussels as capital of Europe

Brussels is not the city that most Europeans would choose as their capital. It shows the mean materialist side of the continent, boring and bourgeois … Yet Brussels is now the nearest thing to a capital of Europe. The trend began with the setting up of the common market and Euratom, when the parliamentarians of Europe couldn't decide between Strasbourg and Paris, and had to settle for Brussels; the common market headquarters attracted armies of diplomats, delegations, pressure groups, trade associations, trade unionists, supplicants, intriguers. In the sixties more and more big companies, particularly American, moved their headquarters to Brussels, often from Geneva or London; their names stare out from the new parts of the cities. Then in 1967 a whole new contingent arrived when NATO and SHAPE were chucked out of France by De Gaulle, bringing another army of hangers-on.

(Sampson 1968: 28–9)

Brussels is the city that for many embodies – for ill, usually – the reality of present-day European integration. The headquarters of the European Commission, the Berlaymont, evokes the notion of faceless civil servants taking decisions that effect the quality of life of all Europeans. Like many cities across Europe – Lisbon, London, Paris, Madrid, Berlin, Rome – Brussels was once a European imperial capital with the bombastic architecture and social histories that are common to the genre (see Driver and Gilbert (1999) on the legacies and 'performance' of European imperialism on its cities). The existence of the EU has brought affluence. In 1998, the EU employed around 20,000 people in the city, with its dependent and related services employing an additional 38,000 people. This has a clear multiplier effect: 'for each EU job, another two are created in EU-dependent and EU-related institutions. For every Euro spent by the EU, another three are spent by dependent and related organizations' (Baeten 2001: 121–2). Given the additional exis-tence of NATO, international headquarters, and international schools, Brussels is one of the leading business centres of Europe. Unlike other national capitals that have sought to promote culture efflorescence (even while built on military and colonial domination) Brussels is defined by its wealth, and its location at the heart of the richest regions in Europe, even if the 'middle' is now shifting eastwards.

In many ways, then, we can see Brussels as being a kind of 'laboratory', 'a unique, multi-scaled place [that] ... urbanizes the very notion of a "unified Europe" through its cosmopolitan population, international service economy and government institutions' (Baeten 2001: 118). There are two issues here: first, should Brussels act for the EU as a capital city does for a nation state, yet in a post-national Europe?; and second, what is the actual role of the city as a home to the human agents of European consciousness, as a city of bureaucracy? And, relatedly, how is the city 'othered' by the likes of British Eurosceptics, who identify the city with all that is evil about the EU?

> Brussels – synonym for the EU, and self-styled capital of Europe – perfectly illustrates the gap between elites and life 'underneath, where the people are', as Pope John Paul II memorably put it. Brussels is a place where highly sophisticated, multilingual men and women from the most diverse backgrounds – a French technocrat, a former governor of Hong Kong, a one-time student opponent of Franco – try to reconcile national interests and national ways of thinking with the pursuit of a larger, common interest. It is also the capital of a country that has almost fallen apart in the conflict between its French-speaking and Flemish (ie. Dutch)-speaking parts, Wallonia and Flanders ... Romano Prodi said that Belgium 'might be considered as a model for Europe'. Indeed.
>
> (Garton Ash 2001: 64)

As I discuss in the next chapter, the symbolism of nation-states is often nurtured most care-fully in the capital city. As a site of pomp and circumstance and state power, as a place of polit-ical protest and as a location of media and government functions, what takes place in the capital is often of utmost importance for the functioning of the broader territory. Yet Brussels is for many a shell, floating free from its own nation-state (Hertmans 2001; Laermans 1999).

Figure 3 *Berlaymont building, Brussels, home of the European Commission. © Audiovisual Library European Commission.*

Could it be, then, that Brussels itself is that city closest to giving answers to a possible European future, a cosmopolitan ghetto, an island in a sea of warring ethnic minorities? As a centre for the grand process of European integration, and with 15 member states each bringing a coterie of civil servants, politicians, and staffers with them, it is obvious that Brussels will be a highly multicultural place. In the EU district, over 40 per cent of the population is non-Belgian. Yet this core is surrounded by a bilingual city-region that represents the rise of the 'particular', of a celebration of ethnic exclusion and distinctiveness (Baeten 2001: 124).

Yet in the midst of this affluence (Brussels is one of the few places in Europe that can rely heavily on public funding to sustain its wealth), it is clear that there are many who are not benefiting. High rates of youth unemployment, especially among ethnic minorities, are a microcosm of the urban economy in Europe as a whole. The 1990s saw a number of ethnically-driven street riots, reflecting the scapegoating of many of these youngsters for urban problems. Baeten (2001) invokes Neil Smith's (1996) notion of the 'revanchist city', where the middle classes empower the police to take back the city streets from the 'dangerous' classes. Interestingly, the local state's attraction of the Eurocrats to live in the city core – a 'European multiculturalisation' – is used 'to re-install a more homogeneous western white middle-class comprehensible urban order' (Baeten 2001: 125).

Whether one accepts this or not, the impact of 'Brussels' as an agent of standardisation is legendary:

Other than ennui, the most important product of Brussels was uniformity. Whether you were there to discuss permissible banana curvatures or the number of hours animals might be carried in a closed truck before they were slaughtered, the European day was the same. You arrived early, after a flight, entering an airless room with translation boxes. Many people were there – too many for real business – and the morning was therefore spent in procedural wrangling. By midday, the British and the French were at war – *bien entendu* – but stomachs were rumbling. At lunch a delicacy was consumed – the white truffle, perhaps. But the afternoon mood was no better and the assembly began to fragment, North against South, Protestant against Catholic and, by the end, everyone else against the Danes and the British. At six o'clock, however, when dusk was turning the windows grey-violet, the bureaucrats moved in to negotiate a resolution. No one was happy with the outcome, but everyone was free to cast blame and take comfort in the memory of the white truffle.

(Fraser 1998: 27).

Unlike the United States, with which the fledgling EU is often compared, the enormous linguistic, economic, and cultural diversity has proved a serious challenge to proponents of integration. Above all, it is the scenes of bureaucratic stalemate portrayed above (albeit slightly facetiously) that has become the stock in trade of nationalist journalists and politicians, as well as sceptical Europhiles. As Shore (2000, Chapter 6) describes, there is frequently used French term in the EU – *engrenage*, meaning 'gearing' or, figuratively, as being 'caught up in the system' – that captures the place effect that Brussels possesses. *Engrenage* is often used to describe the socialisation of national civil servants and politicians in the workings of the transnational EU Commission, such as the technical policy discourses used: '*Engrenage* is therefore best understood as a "political technology" or administrative instrument designed to forge European consciousness and European identity among those policy professionals who operate above the level of the nation-state' (Shore 2000: 148).

In this context, making Brussels synonymous with bureaucracy has some validity. In the UK, this discourse where 'Brussels' is 'othered' intensified towards the end of the 1990s, particularly in right-wing newspapers such as the *Sun* and *Daily Telegraph,* which regularly printed a series of – usually half-true – accusations against European directives. As a riposte, the European Commission's UK press office released a 'glossary of Eurosceptic beliefs': 'an exposé of misunderstanding' (www.cec.org.uk/press/glossary.htm) which detailed the range of accusations levelled against the Commission. The myth with the greatest impact is the 'straight banana' controversy, in which the *Sun* reported the apparent stipulation that 'bananas must not be excessively curved' (4 March 1998: 6). The Commission response was as follows:

Bananas are classified according to quality and size for international trade. Individual governments and the industry have in the past had their own standards and the latter, in particular, have been very stringent. The European Commission was asked by national agricultural ministers and the industry to draft legislation for ministers to agree.

(www.cec.org.uk/press/glossary.htm)

Such a response hardly scotches the myth, but does reveal one thing – that it is the demands of harmonisation on the basis of international trade that is fuelling such standards, and these are being driven by national politicians in the Council of Ministers (which steers the Commission to a degree).

So, as the capital of the New Europe Brussels has multiple roles. But it remains un-inspiring, hated, even. As a shorthand for bureaucracy, as a signifier for the threat to national sovereignty, 'Brussels' is a powerful symbol. As co-ordinating centre for the European project, it remains to inspire affection. Timothy Garton Ash has described it like this:

> On close inspection, Brussels does have a quiet, private drama: individuals of diverse and historically opposed nationalities daily struggling to go beyond national interests and linguistically anchored national ways of thinking – in short, to be that mysterious thing, European. But it has no public drama. The nearest one gets to political theater is at important summits like Nice, but they are largely reported as international diplomatic fencing matches.
>
> (Garton Ash 2001: 66).

In recognition of this, in 2002 the Commission president, Romano Prodi, organised a summit of some of Europe's most prominent thinkers and cultural professionals to generate debate about the future of Brussels and the EU (Wise 2002). These included Umberto Eco, the Italian author of *The Name of the Rose*, the Basque Juan Vidarte, director of the Bilbao Guggenheim, the Dutch architect Rem Koolhaas (designer of the Lille high-speed train station, and champion of 'shopping' architecture such as Prada's Manhattan flagship store, among other things), the historian and ex-foreign minister of Poland, Bronislaw Geremek, and Gérard Mortier, former director of the Salzburg (classical music) festival. Their responses were diverse: Koolhaas advocated a barcode to act as symbol for the EU, an apparent allusion to the consumerist dimension of European integration; Geremek favoured a floating, non-fixed capital, a reference to the medieval, pre-national form of political sovereignty; and Eco, as a statement against the architectural pomp that capital cities often inspire, argued that the Manneken-Pis (the kitsch statue of a urinating small boy that sits in the centre of Brussels) was the most appropriate symbol. Yet the most interesting fact about the summit was Prodi's feeling that there was a need for such a rethink: as I discuss in Chapter 2, Europe is in a competitive market of jostling national 'brands' of political and cultural identities (Van Ham 2002).

Europe and its others: America, Islam, and enlargement

> We should be concerned about the way in which the simple idea of 'imagined community' has come to prevail in contemporary cultural debates. What was once, indeed, a fertile and productive concept has now, I believe, come to inhibit our further understanding of collective cultural experience. We should not let ourselves become comfortable with the idea of 'imagined community'. It draws us into the contemplation of the need for, and entitlement to, shared community, and, by the same token, away from the more difficult question of our cultural responsibilities beyond 'our' cultural world.
>
> (Robins 1999: 273)

What Robins fears, it seems, is the closing of ranks that come about in an imagined community. It identifies a space, a shared set of values, and then excludes those who do not belong. So, the emergence of a dual re-imagining of, on the one hand, a European super-state with associated cultural and political forms, and on the other a return to the small, 'homeland' of particularist identity is likely to raise conflict and raise boundaries, just at a time when many hope they are on the wane. As such:

> The discourse of Euro-culture is significant: it is that of cohesion, integration, union, security ... European culture is imagined in terms of an idealized wholeness and plenitude, and European identity in terms of boundedness and containment ... Imagined in this sense, the community is always – eternally and inherently – fated to anxiety.
>
> (Robins 1999: 273).

There is two important sources of anxiety here: the idea of an external enemy, and the idea of the 'resident Other'. First, if Europe is defined as a community then it implies the existence of alternatives, namely the economically and culturally powerful US, a vaguely-defined Islamic threat, and a pre-democratic 'Wild East' (Van Ham 2001) in the countries of the disintegrated Soviet bloc. Second, there is the idea of members of these outside groups (particularly the last two) as being internal threats to a European identity:

> With its millions of (illegal) immigrants and denizens, Europe has many minority groups within its boundaries that are perceived to injure its cultural and social cohesion. The resident Other is not comfortably spatially distant, but often lives across the street and confronts us with different mores, values and practices that challenge hegemonic cultural patterns.
>
> (Van Ham 2001: 195)

So, we have the idea of a bounded culture that is perhaps threatened by the airwaves that emanate from Hollywood and CNN, driving an 'anglicisation' of languages. And we have the idea of the 'Other within', where those with different skin colours or religions may disrupt the concept of an 'imagined community' that both Europeanists and nationalists promote.

In any discussion of European identity, it is therefore crucial to consider how this territorial formation that we now may call the European Union relates to the 'outside world', both in a contemporary sense and historically. First, there is the fact that Europe itself has contributed – rarely pacifically – to the cultural formation of other parts of the world, whether it be the British in India, the Italians in North Africa, the Spanish in South and Central America, the Belgians in the Congo. Here, there is a sense in which Europe can only be understood from a post-colonial perspective. Second, in-coming cultural flows are complex, and external ideas are often received enthusiastically and altered to suit the specific conditions of local cultures, in a process of 'indigenisation' (Appadurai 1990). Third, we now live in a time when many corporations are now transnational, severed from the countries in which they originated to all intents and purposes. The products they sell, their executives, their marketing campaigns may be as European as American, or indeed Japanese, or indeed blend into a corporate 'global culture'. Where, in this complex picture, do we locate Europe?

America, Europe, and globalisation

> Over the decades, the words changed. *Americanization, homogenization, coca-colonization, media imperialism, global hegemony* – at one time or another, critics overseas and in the United States employed all these terms to describe America's domination of the planet through the export of its products, its culture, its way of life … Implicit in almost all the attacks on America's 'cultural imperialism' was the assumption that people in Europe and elsewhere were little more than receptacles, mindlessly ingesting and internalizing the messages of the American media. According to its opponents, mass culture converted audiences into a collection of zombies, docile and passive, too drugged to discriminate between art and trash, too hypnotized to switch off the television set or get off the information highway.
>
> (Pells 1997: 278–9)

It is undeniable that for many Europeans, globalisation is often associated with Americanisation. Ever since the Marshall Plan unleashed the full force of corporate America on a war-torn European economy, there has been a strong fear of invasion, of cultural – replacing military – imperialism that would undermine the essences of national and local belonging. Perhaps the greatest fear is not of Americana itself – many European have marvelled at the exciting cultural products that have crossed the Atlantic, from Coca-Cola to *Friends* to Corn Flakes – but rather of a standardisation, a homogenisation, that places are becoming increasingly similar or losing their distinctiveness, that local commodities are being priced out or superseded. Here, corporations seek to maximise economies of scale in anything from burgers to soap operas by penetrating national markets. At face value, many of the most celebrated (or notorious) examples of this have been perpetrated by American corporations.

However, the identification of America as other, or as enemy (cultural, if not military), has to be qualified in three ways. First, Americans themselves are subject to corporate domination.

> Americans are as ambivalent about what it means to be modern, computerized, and technologically sophisticated – attributes synonymous with the American way of life – as are Europeans. The fear of losing one's unique cultural heritage as one becomes an affluent consumer of America's goods and services, movies, and mass circulation magazines is as strong in the United States as it is in Europe.
>
> (Pells 1997: xvi)

Second, American culture is willingly indigenised by many individuals and social groups. During the 1930s, many Italians sought escape from fascist 'Italianisation' by immersion in American film and fiction. In Germany, jazz became a dominant and early form of mass popular music (to the express hostility of Hitler). According to Pells:

> When Europeans contemplated the 'culture' of the United States, they were not thinking about America's postwar leadership in science, literature, painting, or architecture, as officers at the State Department and the U.S. Information Agency would have preferred. For Europeans in the 1940s and 1950s, even more than for their predecessors in the 1920s, American culture meant movies, jazz, rock and roll, newspapers, mass-circulation magazines, advertising, comic strips, and ultimately television. This was a culture created not for the patricians but for the common folk.
>
> (Pells 1997: 204).

The burgeoning youth culture of post-war Europe seized upon the notion of individual freedom from traditional values communicated through Hollywood, either as rebellion (James Dean, Marlon Brando) or sexual glamour (Marilyn Monroe). Third, it could be argued that European culture (understood as a bundle of nationally-produced artefacts) and people have had a significant impact on the US itself. Here, the 'old world' of exodus and diaspora has had a major part to play in the development of what we know as American culture, whether it be in the mafia, in modernist architecture, in philanthropy, in art or in science.

Nonetheless, one of the most prevalent ways of framing this issue has been to pose it as an American cultural invasion. Malcolm Bradbury has put it like this:

> Seen from the Eastern flank, from Central Europe or Russia, it is, just for the moment, not too hard to know what Europe is. It's the West, the free market – which in fact comes quite expensive. It is Berlin and Milan and Vienna bursting with European, American and Japanese consumer goods. It is the Mercedes and Fiat, the icebox and the CD player, the ecology movement and gender wars, the Beatles and Madonna . . . It is those vivid European cities that have restored their historic past with a clinical cultural nostalgia, yet have not failed to take on McDonald's, the shopping mall, and all the high-rise, skyscraper *chic* of the future . . . And, like America itself, Europe is also a network of signs, satellite systems and communications, multiplying electronic images and electronic rates of exchange. Ever since Marxism failed in Eastern Europe, any clear ideological alternative to these particular (American?) patterns and means of modernization, any different account of social progress or evolutionary history, has for the moment died.
>
> (Bradbury 1996: 479)

It is interesting to note this final ambivalence about whether there is a distinctive European 'culture', for as Bradbury notes there is good reason for seeing Europe as fully modernised *yet at the same time* keen to preserve the 'clinical cultural nostalgia' for a past bundle of cultural images, from Mozart to Bernini to Erasmus. This fear of a loss of the past, a loss of cultural distinctiveness, perhaps fuels some of the most important sectors of the modern economy: tourism, advertising, and audiovisual entertainment. This is all wrapped into a capitalist marketing dream, where places and the past become commodified.

And so, Europe and America have a complex relationship. Ellwood (2000) provides a comparative analysis of why and how different European countries, and different groups within these countries, attack America. He writes:

> The true usefulness of 'Anti-Americanism' as a category of thought or behaviour lies surely in its catch-all nature. Conveniently but misleadingly, it hides the important distinctions between those intent on attacking America the nation, the government, the foreign policy; those who find repugnant whatever or whoever is American: the way of life, the symbols, objects, products and people; and the critics of Americanism, those who reject the explicit values and ideals of the United States in their distinctive normative form.
>
> (Ellwood 2000: 27)

Islam and Europe

While the US has come to stand for uncontrolled capitalism, military expansionism and radical individualism, Europe's second major 'other' has to be Islam. While the events of September 11 hardly served to dissipate this, the construction of Islam as the rival of Europe has centuries of history (Rietbergen 1998). It has been suggested that the demise of the USSR and Warsaw Pact has created the need for an alternative external enemy to mobilise and solidify the West's political project.

Yet as Yasmin Alibhai-Brown (2001) has noted, the 'othering' of Muslims in the West should be tempered by two things: first, that Muslims have a long history within Europe, and were often the target of Christian hostility, such as the expulsion of the Moors from southern Spain in the fifteenth century; second, that most Muslims have a worldview far removed from the religious fundamentalism represented by the Taliban, for example. With 17 million Muslims living in the European Union, making Islam the second largest religion in the EU, she argues that a more far-reaching reassessment of European Muslim and non-Muslim identity is required that 'cannot be based on "celebrating diversity" or "tolerance" or other such anodyne concepts' (p. 214). There are two issues to consider here: the nature of Islam within European society; and the position of Islam within EU foreign and cultural policy.

First, Islam is now a fundamental part of European society, with approximately 15 million Muslims out of a total population of 378 million, around 4–5 per cent (Cameron 2002: 258). As well as being the second most important religion, it has also been an important milieu for the development of Islamic thought. As Esposito (2002: 251) argues, it has been the more open, tolerant nature of European political and legal systems that have 'freed Muslim scholars and students to engage in a rigorous process of reinterpretation of Islamic sources and legal reform' by contrast with some of the authoritarian regimes in the Muslim world that restrict freedom of speech (Iraq, for example), allowing an interesting two-way process of thought. Thus some Muslim countries have funded mosques, schools, and various institutions to promote particular variants of Islam in the West, and simultaneously European-resident or diaspora Muslims may return to their native countries and influence thought and interpretation there. As such:

> Only a few decades ago, it was accurate to talk about Islam and the West. Something that was unthinkable, or even unimaginable, only a few short years ago is now a reality. Today, the cities and learning centers of the Muslim world are not only in Cairo, Damascus, Islamabad, and Kuala Lumpur but also London, Manchester, Paris, Marseilles, Amsterdam, Antwerp, New York, Detroit and Los Angeles.
>
> (Esposito 2002: 246)

The importance of Muslims to major European urban centres leads on to a second issue, the relationship of the EU as an institution to the Muslim world. Here, the enlargement of the EU since the 1980s has meant that it is now territorially very close to the Middle East and North Africa, the Muslim heartland. Furthermore, its prosperity depends on maintaining trade and energy supply relations with stable governments, as well as accommodating the demands of Muslim citizens. This partially explains the EU's (flawed) desire to intervene to end conflict in Bosnia and Kosovo (Cameron 2002). It also explains the growing interest of the EU in supporting the fledgling Palestinian Authority, a counter-balance to US policy in the region, and in the Euro-Mediterranean Partnership (the 'Barcelona process') launched in 1995 to promote greater trade, cultural, and political co-operation with the countries on the Southern and Eastern shores of the Mediterranean (Cameron 2002).

These more subterranean, even abstract, ideas have been given more concrete form with the attempts of Turkey to join the EU. Turkey offers an important bridge between the Southeastern borders of the Union, and applied for membership in 1987, yet was bypassed in 1997 by a list of no fewer than nine other applicants, many from the post-communist East and Central Europe. Although talks reopened in 1999, and while there were apparently clear reasons for the lukewarm reception – their treatment of the Kurds and tensions with Greece, particularly – the important subtext to the discussion was Turkey's substantial Muslim population, which would significantly alter any claim to Europe as having a coherent identity based on shared Christian traditions (Van Ham 2001: 209–14).

Enlargement: Mitteleuropa, Russia, and Balkanisation

> The idea that East Europe is a product of its own past is, then, self-serving for the West. It excuses the devastation wrought by the major powers, by pointing to a native 'East European-ness'.
>
> (Burgess 1997: 107)

The attitude of many Western politicians (both European and American) to the countries of Eastern and Central Europe before 1989 was clear. Here, plucky nations were being subordinated to the mighty Soviet (read Russian) military and ideological machine. Every effort was made to encourage them to seek freedom (read consumerist lifestyles). After 1989, however, things changed. The West was not quite as enthusiastic about embracing these countries into the European club. Instead, they were constructed as new markets, as new lands to invest in, to expand European and American capitalism eastwards. In 1998, accession negotiations were opened with Estonia, Cyprus, the Czech Republic, Slovenia, Hungary, and Poland, the countries seen to be best-positioned to fulfil the membership conditions of the EU. In 2000, accession negotiations were opened with Latvia, Lithuania, Bulgaria, Romania, and Slovakia, who now appear likely to also be able to join the Union. The idea that there is a single East and Central Europe (ECE) is, logically, contested in various ways (Van Ham 2001).

Smith's (2002) discussion of this rescripting – effectively a whole new discourse, a whole new way of talking about the East – highlights a number of aspects which reflect the West's geo-economic power. Each of these is bound up with the notion of *transition,* the leaving behind of the norms of communist society, and their replacement with the values of liberal democracy. First, the 1989 collapse of state communism was heralded by the West as an opportunity to expand the neo-liberal transition in the US and Western Europe to the East. This reached its logical extension with the creation of a permanent financial institution to oversee the transition from centralised and nationalised state economies to a more market-led economy, the launch of the European Bank for Reconstruction and Development (EBRD) in 1990. The indicators used by Western governments and institutions such as the EBRD and IMF are tied exclusively to plain logics of profitability and market success, rather than indicators of, say, social goals such as literacy, education, health, working conditions, and so on (Smith 2002: 651–6). Second, relatedly, there is the role of 'financial surveillance' of new economic frontiers. These 'emerging' economies are tracked by databases run by, for example, Standard and Poor, to provide knowledge for investors. Here, emerging and frontier markets are tracked, implying a 'frontier' myth similar to that in American political history, 'rolling back' the wilderness (and making the world safe for capitalism) (Smith 2002: 657–9). Third, there is a naturalisation of uneven development, where certain leading neo-liberal academics have argued that *physical* geographical traits (eg whether a country is coastal or land-locked) may have impacted upon 'native' attitudes to capitalism and trade (Smith 2002: 659–60; Burgess 1997).

These three Western discourses – tied to concrete policy strategies such as the EU's Phare programme of technology transfer – have shaped the question of *enlargement,* the debate over whether the EU should include countries from the ECE region as members, given the disparities between standards of living and market strength. In this context, it is argued that countries such as Hungary, Poland, and the Czech Republic are 'othering' the countries to their East, creating a clear distinction between the ECE countries from the region of South Eastern Europe (SEE). These countries – Albania, Bosnia-Herzegovina,

Bulgaria, Croatia, the Federal Republic of Yugoslavia, FYR Macedonia, and Romania – are defined by the World Bank as having shared characteristics of lack of progress towards capitalist normality, by contrast with the post-communist Baltic and Central European economies. Furthermore, the West's financial expansion desires are also linked to a fear of crime, immigration, and military conflict (perhaps requiring expensive and unpopular intervention).

The existence of these 'othered' nations is often discussed in terms that recall the tortuous geopolitics of the early twentieth century, where major European powers were ultimately sucked into warfare as a result of their adventures in the 'Balkans'. As such, the ethnic conflicts that engulfed the region identified above – the area centred around the ex-Yugoslavia – were often reported in terms that emphasised the 'primordial', or impossible, nature of political (in)stability. Even before the outbreak of the conflicts of the 1990s, 'Balkanisation' came to stand as a *general* metaphor for the fragmentation of small, deeply nationalist states within a larger federation, following a deep-seated stigmatisation of the 'East' by the 'West' throughout twentieth century geopolitical thought and policy (Burgess 1997, Todorova 1997).

This stance often appeared in the recreation (or reimagining, more properly) of *Mitteleuropa,* an idea with a complex history (Hagen 2003; Heffernan 1998: 71–6, 220–2). Its original incarnation, a European customs union with Germany and German culture at its core, was pursued in numerous forms, both socialist and conservative, in the first decades of the twentieth century. In the post-1989 context of Europe, a number of writers – notably the exiled Czech novelist Milan Kundera – called for its return, invoking the geographic romance or nostalgia of the region, its democratic political traditions, its cultural innovation, its place within European modernism. Furthermore, it had a distinctive purchase on the hitherto technocratic idea of a Europe based on industrial agriculture, a new motorway network, and free trade. For example, in his discussion of *Danube* (1990), Malcolm Bradbury notes how the author Claudio Magris evokes:

> the sensation of travelling eastward across Europe, from the world of the Rhine to the world of the Danube, from Europe to non-Europe [sic.] . . . his tale of travel takes in Freud and Wittgenstein, Kafka and Canetti and Lukács, the great European modernist thinkers of the difficult way down the river, and it likewise takes in the great and confused cities of Modernism – Vienna, Budapest, and not too far away Prague – where the modern thinking was done.
>
> (Bradbury 1996: 478)

While the strengthening of the *Mitteleuropa* idea is often seen as politically progressive, given the emphasis on a construction of democratic rights and a culture of citizenship, in some of its manifestations it can be exclusionary. Some supporters of a Central European idea were keen to draw a borderline between themselves and those (to the East, generally) they saw as being less civilised. This 'backwardness project' means that the East–West division persists, but has been actively pushed eastwards. As Kürti puts it:

> There are now countries – namely Poland, the Czech Republic, Hungary, Croatia and Slovenia – which proudly boast the most distance from their communist past and identify themselves as Central, or East-Central European, while in contrast, they view other states – Slovakia, Bulgaria, Romania, Serbia, Albania, Ukraine and Russia – as East Europeans proper, whose national heritage includes a much more serious dosage of communism and, thus, backwardness.
>
> (Kürti 1997: 39)

Within this reading then, historians and politicians and media in these 'central' European countries are constructing a dichotomy or set of dualisms that effectively put up another border. Central Europe is seen as having far-reaching democratic reforms, the embrace of a market economy, and an increasingly sophisticated consumer economy. Eastern Europe is characterised by only partial democratic reform, a high degree of state involvement in the economy, and a poorly developed consumer economy (Kürti 1997: 44). Wrapped within all these dualisms is the sense of belonging to Europe, that being European means being 'civilised', 'democratic', and so on. Thus many in Poland and Czechoslovakia used the idea of a 'return' to Europe, in other words the re-embrace of a history and cultural identity that had been torn away first by the Nazis, and then by the Soviets.

The position of Russia in this is very important. In one of his later acts as Soviet leader, Mikhail Gorbachev floated the idea of a 'Common European Home,' a statement that reversed the Cold War logic of clearly demarcated European 'spheres of influence'. Gorbachev invoked *Mitteleuropa*, but argued for the significance of Russia as having a clear European heritage (an idea first pursued by Peter the Great). Events would subsequently overtake Gorbachev's attempt at a managed reform of the Soviet Union, and its disintegration (along with that of Yugoslavia) raised all sorts of questions about democracy, culture and identity in the East. Van Ham (2001) summarises this as follows:

> The civil wars in Bosnia, Kosovo and Chechnya have raised Europe's awareness that new, fuzzy dangers are again coming from the east. This is the source of nuclear suitcases and leaking submarines, violent Mafia gangs, loony radical politicians with wild expansionist designs, as well as the manifest crumbling of a sense of law and order.
>
> (Van Ham 2001: 207)

As such, the question of enlargement remains one of fundamental concern to any idea of a European identity.

Conclusion

In this chapter, I have tried to describe Europe as a construction. In other words, it is something that operates discursively and symbolically, talked into being by politicians, bureaucrats, and ordinary people, rather than simply being a simple description of the final state of an integration project. This has fundamental implications for whether we are Eurosceptic, enthusiastic, or simply confused. To conclude I want to make three points.

First, Europe can be understood visually and experientially, in the landscapes we view, the cities we visit or live in, food we eat or cars we drive. It is a clearly geographical project, in the sense that it invites debate on place identity and belonging. This is important whether we the New Europe as a set of fragments, as in Enzensberger's (1989) *Europe, Europe,* or whether as a set of traces on landscapes, as Jan Morris (1998) has evoked – sometimes rather cloyingly, but sometimes rather charmingly – or as a more abstract set of networks and social processes, as Jönsson et al. (2000) have sketched in *Organizing European Space.* Here, to reiterate a point I made in the introduction, social scientists have to be aware of the assumptions they make and the standpoint they take when talking of Europe.

Second, relatedly, some are tempted to view Europe as a coherent entity with borders that are closed, and an identifiable culture in common. As Massey (1995) has argued:

> Those claims for European identity which look set to become the dominant ones
> generally evoke a continuous and singular history, an uninterrupted progress to the
> present, and it is by and large an internal one. They seek the European character *within,*
> denying its constant external connections: the fact of the construction of the local
> character of Europe through its constant association with the global, whether invasions
> for the vast openness of the East in the distant past, the initial connections of
> mercantilism and imperialism (from the China Seas to North Africa to the Caribbean), or
> the physical presence of 'ethnic minorities' within its borders now. If the 'outside world' is
> recognised at all in this approach to place-definition it is through negative
> counterposition (this place is *not* Islamic, not part of the Muslim world), rather than
> through positive interrelation.

This, as I described above, can be conceptualised as Europe against its Others, be this America, Russia, or Islam, and – as with the nation-state – seeks legitimacy through the construction of a common foe, real or imagined. This idea will be challenged throughout the book, with consideration of the idea of transnationalism and the nature of borders at various points. Alternatively, defenders of the nation-state may construct 'Europe' as its other, despite the fact that it is national politicians who have been constructing it.

Third, when looking at where this New Europe may be located most intensely, perhaps it would be worth looking at the intersection of Americanism, national identity, and the East in the 'shock cities' of Central Europe, evoked well by Timothy Garton Ash:

> While Budapest gradually developed into a modern consumer city, starting in the 1970s,
> Prague has emerged from its time warp suddenly and explosively. Instead of the magical
> museum, lovely but decaying, there is colour, noise, action: street performers, traffic jams,
> building works, thousands of young Americans – would-be Hemingways or Scott
> Fitzgeralds – millions of German tourists, betting shops, reserved parking places for
> France Telecom and Mitsubishi Corporation, beggars, junkies, *Spesenritter* of all countries,
> car alarms, trendy bars, gangsters, whores galore, *Bierstüben,* litter, graffiti, video shops and
> Franz Kafka T-shirts.
>
> (Garton Ash 2000: 117).

So, to follow Van Ham (2001) again, it is my standpoint that Europe is messy, lacking order, and certainly lacking any one grand theory – geographical or political – that allows us to grasp it. It is certainly in the interests of the EU elites to push forward a single history or political understanding of the project, as much to simplify a tortuous process as to hoodwink Europeans. But in the following chapters I try to highlight the complexity of understanding the New Europe, due on the one hand to the often competing imaginaries of region, nation, and city, and on the other to the fact that Europe is a continent in motion.

2
Europe and the nation

> [The] spatial scaling of the nation operates at a variety of levels. It is present in the televisual space (in the space of the home) which is beamed to a (national) community of viewers, which transmits a host of spatial images, including national ideological landscapes, iconic sites and sites of popular congregation, everyday spatial fixtures and the mundane landscapes of quotidian life. It is facilitated by technologies of mobility which enable people to travel across the nation and experience national signs and the regional distinctions which are identified as being incorporated into the nation ...The elements of national space are linked together to constitute practical and symbolic imaginary geographies which confirm the nation as the pre-eminent spatial entity.
>
> (Edensor 2002: 65–7).

For much of the twentieth century, Europe's nation-states have been the principal mode by which cultural, social, political, and economic life is understood. Yet this dominance is under threat, as the various processes of globalisation are played out. For James Anderson (1996: 133), 'the ground is shifting under established political institutions, practices, and concepts'. The suggestion that we are witnessing the death of the nation-state, as it loses its sovereignty over anything from foreign policy to control of its currency and economy, is persuasive. However:

> In some ways the same old territorial states with their sovereignty defined by the same old borders seem as firmly rooted as ever: tax collectors stop at the border, immigrants are stopped at the same border, and transnational linkages can still be arbitrarily snapped off by independent state powers.
>
> (J. Anderson 1996: 135).

So, the debate over Europe's territorial futures is often excessively simplistic, assuming either the continuing existence of a nation-state led form of European integration (inter-governmentalism, where states agree to pool sovereignty) or a federalisation model that simultaneously enhances the power of the regions and the EU at the expense of national autonomy.

However, 'many cultural, social, moral and military issues are dealt with only tangentially [by the EU] if at all; political parties are still almost entirely national in character; and even EU elections are often more about national rather than European issues' (J. Anderson 1996). One thing is clear, it would seem: nation-states are not ignorant of the EU, and nor are they dissolving themselves within it. Instead, there is a more complex political process of repositioning, remapping, and rebranding of the nation-state often led by political parties and their leaders. They will include responses to external military forces, cultural flows and terrorist networks, new forms of national political projects, both progressive and reactionary, and the transformation of 'national' landscapes and cityscapes by new infrastructural and architectural projects. Globalisation and Europeanisation may be simultaneously welcomed and fiercely resisted by different groups in the same national society. I begin this chapter, however, by setting out differing ways in which the 'nation' and its symbolic and imaginary geographies can be understood and represented.

Representing the nation

> Myths and symbols provide key components of understanding and directing change . . .
> [The nation's] myths and symbols orient its members to fundamental values – to homeland, the defence of 'irreplaceable cultural values', and to the freedom of its people. Images establish a common bond between leaders and led and legitimise even in 'normal times' difficult decisions – the imposition of heavier taxes in economic difficulties, the redistribution of resources between richer and poorer regions or classes.
>
> (Hutchinson 2003: 49)

Van Ham (2001: 60) makes a useful distinction between views of nationalism as, on the one hand, 'primordial', and, on the other, 'ephemeral'. Here, the idea of primordialism means that nations are somehow natural, and 'stresses the differences from 'others', who are considered outsiders and cannot be converted or even adopted . . . This implies that such strangers are often perceived as a threat to the natural homogeneity of the collectivity' (p. 60). By contrast, a view based on ephemerality sees the concept of nationalism as going through different historical phases. First, a premodern, preindustrial society was based very strongly on locality and face-to-face contact. With the coming of industrial society and increasing human mobility, states had to 'forge' a sense of communal identity and collective understanding, through the 'invention of tradition'. In the latest stage (and this is of course a very crude categorisation), there is a sense in which new communications technology and mass consumerism may be homogenising or under-mining national identity.

> This is the postmodern cosmopolitan culture that some consider the pinnacle, others the nadir, of human progress. It is a pastiche of cultures, rather than based on one, specific culture. It is eclectic in nature, disinterested in place and time, unconcerned about ethnic or national origins and blissfully ignorant of history.
>
> (Van Ham 2001: 63)

Just as Europeanisation is a live process, and just as globalisation brings with it all sorts of disorienting, 'placeless' aspects to identities, so nationalism has not remained static. In many of the examples discussed below, it is clear that all the major nation-states of Europe have encountered problems in their future direction and their attitude to the past, challenges posed in many cases by the intensification of European integration. It is not always clear that these are 'nationalist' responses, however, as Michael Billig argues:

> In both popular and academic writing, nationalism is associated with those who struggle to create new states or with extreme right-wing politics. According to customary usage, George Bush [sr] is not a nationalist; but separatists in Quebec or Brittany are; so are the leaders of extreme right-wing parties such as the Front National in France; and so, too, are the Serbian guerrillas, killing in the cause of extending the homeland's borders.
>
> (Billig 1995: 5)

Yet:

> Gaps in political language are rarely innocent. The case of 'nationalism' is no exception. By being semantically restricted to small sizes and exotic colours, 'nationalism' becomes identified as a problem: it occurs 'there' on the periphery, not 'here' at the centre. The separatists, the fascists, the guerrillas are the problem of nationalism ... surely, there is a distinction between the flag waved by Serbian ethnic cleansers and that hanging unobtrusively outside the US post office; or between the policy of the Front National and the support given by the leader of the opposition to the British government's Falklands policy. For this reason, the term *banal nationalism* is introduced to cover the ideological habits which enable the established nations of the West to be reproduced.
>
> (Billig 1995: 6)

Billig looks at numerous examples – especially of the mass media and political speech-making – of this banal nationalism. His concerns echo those of Benedict Anderson, whose 'imagined community' idea is often boiled into a single statement: what makes people living in the same national territory go into war for the sake of those they may never know? Or less dramatically, we could ask: what makes people seize the idea of nationhood as an marker, why is it so important to proclaim oneself Scottish, English, or French?

The actual process of national imagining is, therefore, seen to be fundamental, as nationalism is often held to be somehow irrational (i.e. few, if any, nations have an un-complicated and pristine history, or a purity of origins). Yet where do we look to discover this imagining? There are four areas to consider.

First, there is the idea of the nation's 'story', which is communicated both formally (e.g. through the educational curriculum) and informally (through myths and folk tales). It is often politically scripted (the rhetoric used, for example, by Conservatives in seeing Britain as an 'island'), or 'a few words suggesting a before-and-after story may suffice for a patriotic narrative: "Italy has awakened" (i.e. Italy was asleep)' (Dickie 1996). The classic institutions of the nation-state, many of which are less than a century old – such as parliaments (and the ability for a nation to decide its own future), coinage (the battle to save the pound), state religion (the role of the Church in Ireland, Italy, and Spain, for example), media (national broadcasting, as in the BBC), and alteration or eulogisation of symbolic landscapes are all central to discourses of and about the nation.

Second, relatedly, there is the use of symbols, be they flags, anthems, national sports teams, or less formal things like foods – the pizza, for example (Dickie 1996). I noted in the last chapter the efforts made by the EU to invent a European image through such techniques, where everything in the above list – except the sports teams – has come into existence (and I assume the Euro-banana might be a good example here). As an example of how nations might be motivated by such things, consider the victory of the French national football team in the World Cup of 1998. In the tournament, the French team had performed surprisingly well and had beaten the favourites, Brazil, by 3–0 in the final in Paris. Remarkably, its squad consisted almost entirely of players who were born either entirely outside France (Karembeu in New Caledonia and Desailly in Ghana, for example) or with parents or grandparents born outside metropolitan France (Henry, Vieira, Djorkaeff, Trezeguet had at least one parent from Guadeloupe, Senegal, Armenia, and Argentina respectively). Most exciting of all, for many, was the fact that the star player of team and tournament – Zinidine Zidane – grew up the son of Algerian Kabyle parentage in a very poor district of Marseille. Here, after years where right-wing demagogues such as Jean-Marie Le Pen had tried to present immigrants as essentially a problem to France, so a new vision of a multicultural nation appeared to be celebrated by the multitudes that crammed central Paris on the night of French victory. *Les Bleus* – as the team is collectively known – thus stood, even if only temporarily, as an important mobilisation of a nation built around difference and mixed cultural and ethnic roots (Marks 1999).

Third, very importantly for our purposes here, is the idea of the nation as a geographical space 'filled with richly connotative land- or city- scapes'. As Edensor notes:

> It is difficult to mention a nation without conjuring up a particular rural landscape (often with particular kinds of people carrying out certain actions). Ireland has become synonymous with the West coast (see Nash 1993), Argentina is inevitably linked with images of the pampas: gauchos riding across the grasslands, Morocco is associated with palm trees, oases and shapely dunescapes, and the Netherlands with a flat patchwork of polders and drainage ditches . . . These specific landscapes are selective shorthand for these nations, synecdoches through which they are recognised globally. But they are also loaded with symbolic values and stand for national virtues, for the forging of the nation out of adversity, or the shaping of its geography out of nature whether conceived as beneficent, tamed, or harnessed.
>
> (Edesor 2002: 39–40)

Interestingly, the idea of similarly connotative cityscapes may jar with this rural vision. As Edensor continues, the idea of Englishness has often been forged out of a myth of a timeless landscape, ignoring the fact that such landscapes are now heavily industrialised, mechanised, and commercialised. However, it could be argued that some nations draw a similarly romantic representation from an idealised *urban* past – Italy, most obviously, is usually associated with the medieval or renaissance architecture of the likes of Florence, Venice, and Rome as much as with a ruralised set of images. Furthermore, as I discuss below, the events of the recent past may still resonate very strongly over debates about national identity. In particular, the decision to move the capital of reunified Germany from Bonn to Berlin – with all the memories of the past that this stirred – provoked an intense (and, for many, healthy) public debate about the nature of Berlin in terms of German identity.

Fourth, there is the idea of the nation and the 'other'. As Fraser (2000) has shown in his travelogue of the far right in Europe, the process of othering is very much alive in 'borderless' Europe, whether in the jack-booted crudities of extreme-right wing movements, or in the smoother, 'respectable' image of the neo-fascist movements of Jörg Haider in Austria and Umberto Bossi in Italy, both of which have gained national office in recent years. While I turn to this issue again in relation to borders in Chapter 6, it is worth noting here that one of the most powerful identity markers of a nationality is by defining it as what one is not . . . 'We're not like the French', for example. As Caryl Phillips (1987) discovers in his journey through Europe, such identities may be on the rise, and at their most prevalent in the most 'civilised' parts of Europe, places with long traditions of democratic government. And it is also worth noting that within the territories of nation-states, there are as virulent battles between the regions and micro-nations of the New Europe over the share of central government funding or the right to cultural autonomy as there are between many nation-states.

The national routine

A further development of the idea of banal nationalism involves a shift in perspective from how the nation is represented, to how it is *performed* or *practised*.

> Performance is a useful metaphor since it allows us to look at the ways in which identities are enacted and reproduced, informing and (re)constructing a sense of collectivity. The notion of performance also foregrounds identity as dynamic; as always in the process of production . . . By extending the analysis to other theatrical concepts we can further explore the meaningful contexts within which such action takes place . . . by conceiving of symbolic sites as *stages*, we can explore *where* identity is dramatised, broadcast, shared and reproduced, how these spaces are shaped to permit particular performances, and how contesting performances orient around both spectacular and everyday sites.
>
> (Edensor 2002: 69)

In his discussion of performance and national identity, Edensor (2002) distinguishes between three forms of everyday, 'mundane' practices which are often done without any conscious reflection yet which are crucial parts of national 'lives'. Of course, these can be over-generalised, and they are – as Edensor recognises – inflected by class, gender, ethnicity, and various other social and individual markers. Yet taken together they provide some of the key differentiating factors that make European nationalities distinctive, an illustration of some of the biggest obstacles to forging a European citizenship.

The first category of interest to Edensor he terms *popular competencies,* by which he means the ability of the citizen to fit into the regulations imposed by the state, but also extends to the kind of skills for 'successful living': 'where to buy certain things, where to seek particular bargains, how to buy theatre tickets, where to enrol at local libraries, where to go and how to worship' (p. 93). The list is almost endless, but the key point is that when we leave the national context and go to a different society, 'we come across a culture full of people who do not do things the way we do them, who draw on different practical resources to accomplish everyday tasks' (p. 93). Crucially, these distinctions determine the ability to relate to other national cultures.

Second, Edensor refers to *embodied habits,* which include modes of walking, sitting, talking, and 'unconscious emotional communication – gestures, smiles and body language' (p. 94). These competencies – the national 'ability' to be good at sex, cooking, dancing, playing football, gardening, are fuel for the stereotypes that solidify national identities, and thus pose the question: is there a *European* way of expression, or a series of national ways of expression?

Third, there are the day to day ways in which societies organise themselves temporally, *synchronised enactions* where 'without recurrent experiences and unreflexive habits there would be no consistency given to experience, no temporal framework within which to make sense of the world' (p. 96). Here, societies may have a more or less set framework for when particular activities should be carried out: loud noise on a Sunday morning is less acceptable than on a Saturday night, rush-hours structure daily living in big cities, lunch and dinner times tend to be uniform within a national society. These time geographies clearly vary across Europe – and again, they may fit into a stereotype (the Spanish siesta, the Italian *passeggiata,* the British 11 pm pub closing time).

These three aspects of everyday performance are, as Edensor demonstrates, fundamental to understanding how national societies are given coherence. Furthermore, my argument in this chapter is that such habits are often constructed within the European context as being *essential* attributes of *other* national cultures, and are thus open to the stereotyping that makes a European identity such a fragile – perhaps illusory – concept. It will remain to be seen how far the (Europeanised) state (broadly defined to include how nation states operate within an overarching European set of policies), may alter its time geographies, its accepted rituals, its social norms, to fit into an integrated European whole.

The European football ritual

If these everyday activities and time routines often pass unremarked in terms of their territorial/national anchorage, in certain 'staged' activities national attributes become part of a commonsense, 'pop' geography and sociology of other countries. Sport is one of the most obvious cases of this. So for example,

> In the case of Boris Becker and Michael Stick, successful German tennis players in the 1980s and 1990s, English tabloid sports pages were full of comments about their

> German 'efficiency' and their likeness to 'ruthless' German 'machines'. Descriptions abounded about their 'armoury', for instance – their 'howitzer' serves. These comments are widely understood in a post-war English context in which German economic performance outperformed the British economy.
>
> (Edensor 2002: 80)

This is interesting for three reasons: first, because the stereotyping of a supposed essential German set of attributes is commonplace in media coverage of major sporting events; second, in these accounts, there is a confusion of bodily performance and attributes with a more general stereotyping of national physical traits; third, sporting styles become a part of a broader commentary on national characteristics which is then fused with other discourses of national performance – in this case, German economic prowess.

So, if sport is a key ritual in the construction of nation, how does this fit into Europe? As I noted in Chapter 1, there is little doubt that one sport dominates a European mass sports culture – football. Here, long-running tournaments at both club and national level have done as much as anything to engender the idea of a 'Europe' in the popular imagination. And here, I suggest, we can see clearly the contradictions involved in claims of a European consciousness:

> Football unites and divides the continent of Europe. The humble (amateur) and not-so-humble (professional) football ground is, even more so than the outlets of a fast-food chain, the one common, even ubiquitous element of Europe's landscape from the Atlantic to the Alps, and from the Mediterranean to the Irish Sea ... However ... when football is mediated ... the accent is placed upon difference, upon that which divides rather than unites, upon that which is distinctive in terms of identity.
>
> (Crolley and Hand 2002: 157)

Crolley and Hand's (2002) survey of how the quality print media in England, France, and Spain cover major footballing events presents a fascinating conclusion: 'similar portrayals of the English, the French, the Germans, the Italians and the Spanish are ... apparent in broadly the same guises across the continent as a whole and are, therefore, indicative of a common European viewpoint on how we all see each other' (p. 162).

The first area that is of interest here concerns bodily performance and physical stature and appearance. So just as different nations have differing gestures and competences, so they have – apparently – differing footballing styles. So, to take but one example, and to echo the commentary on German tennis players above, Crolley and Hand describe how Spanish, French, and English newspapers concurred in persistently describing various German football teams (especially in the 1998 World Cup) as 'machine-like' or automotive (see Crolley and Hand 2002: 49, 98, 151 respectively). Furthermore, *even the German newspapers appear to subscribe to these beliefs,* according to Crolley and Hand's reading of the *Suddendeutsche Zeitung* newspaper. Hence such myths – if regarded positively – can become self-fulfilling (Blain *et al.* 1993).

In many ways, then, the German team is characterised as having a particular style. Yet this is very different from the actual tactical playing styles characterised by, say, the Italian *catenaccio* ('the lock', signifying extremely tight defensive play at the expense of flair) or the Dutch 'total football' of the 1970s (see Lanfranchi and Taylor 2001, Chapter 7, for more details). National teams are often seen as possessing physically stereotyped individuals, either in terms of stature or in behaviour. Thus, the Italian striker Christian Vieri was described on British television as 'big, strong and powerful, not a typical Italian centre forward, more English in style' by Kevin Keegan (in Crolley and Hand 2002: 47).

The second issue of interest in the context of European identity is that there is often a deliberate confusion of sport with broader military, diplomatic, cultural or socio-economic attributes. Again, this is present across the European media, and ranges from the development of Anglo-German footballing rivalry as a complex fusion of post-war distrust, economic rivalry, and grudging respect for apparent German technical and manufacturing prowess (which may explain the machinic metaphors used to describe the Germans) (Crolley and Hand 2002, Chapter 2; see also Downing 2000), to the coverage of the strikingly multiethnic French team's victory over Brazil in the 1998 World Cup.

Third, while the World Cup and the European Championship have long given meaning to international competition, recent years have seen an intensification in the cultural mixing that exists among most European clubs. While Lanfranchi and Taylor (2001) have done much to illustrate that the actual *foundation* of football clubs is illustrative of the long history of transnational migration within and between European nations, there is also a lot of scope for seeing football as a means of understanding European colonial and postcolonial histories.

To summarise, while in Chapter 1 football was identified as being one of the few areas where a relatively 'organic' notion of a distinctive European identity exists, this in many ways is reinforcing the rivalries that exist among near neighbours. Thus national identity becomes most focused at times when collective European activity takes place. And so, as Crolley and Hand (2002: 162) conclude, 'In short, as Europeans, the one thing we can apparently all agree upon is precisely how we allegedly all differ from each other'.

National car cultures

> [The car is] probably the most richly symbolic artefact of the twentieth century [for] besides its iconic status, it has transformed societies in introducing new forms of mobility. Its emplacement in national cultures has also meant that the geographies it has produced, the ways in which it is represented, the various affordances of particular models, the ways in which it is inhabited and driven, the forms of governance which regulate its use, and the role of motor manufacturing companies have claimed a similarly iconic role in national(ist) imaginaries.
>
> (Edensor 2002: 118)

How might cars be implicated in the performing of national identity? First, the car industry has often perceived to be symbolic of the strength of the national economy. The sale of Rolls Royce in 1998 – based on 'a representation of Britain as class-bound and obsessed with status' (Edensor 2002: 124) – to Volkswagen was an important narrative of the relative decline of the British economy vis-à-vis that of Germany. Similarly, Sampson (1968: 116–34) describes the role of the big national mass manufacturers – Volkswagen, Fiat, Renault, British Leyland – in making the icons of mobility for many working and middle class Europeans: 'the comic little Citroën "deux-chevaux" with its hind-wheels sticking up like frog's legs: the Citroën DSs like sleek porcupines; the tiny Fiat "topolini", the "little mice"; and of course the Volkswagens, the timeless and unmistakeable beetles' (p. 118). So, the car as *manufactured object* has given rise to the concepts of Fordism and post-Fordism which have been applied to a more general analysis of the nation-state under capitalism (Sheller and Urry 2000). Along with the leading centres of American car production (Detroit, for example), across Europe the likes of Volkswagen, Fiat, Seat, Skoda, Triumph, BMW, Citroën, became major contributors to national industrial economies. These centres – Turin, Barcelona, Birmingham – also developed strong working class political parties and trades unions.

Second, Sheller and Urry (2000) note that after housing, the car is the major 'object of individual consumption' which generates status, ('speed, home, safety, sexual desire, career success, freedom, family, masculinity,') anthropomorphism, 'by being given names, having rebellious features, being seen to age and so on', and crime and danger, 'theft, speeding, drunk driving, dangerous driving' (p. 738). The European advertising agencies have all but exhausted the first and second categories in their attempt to naturalise cars in the European psyche. In the latter case, think of Grace Kelly and Princess Diana, Patrick Kluivert, films from *The Italian Job* to *À bout de souffle* to *The French Connection*, or the disastrously dangerous roadways used by clubbers such as Madrid-Valencia, the San Antonio road in Ibiza, routes from Rimini to other big cities in Northern Italy, or any number of treacherous routes across Europe. While the car undeniably has a resonance in American cultural life, its symbolism in European culture has not been fully explored.

Third, there is the impact of the 'national motorscape' on the actual territory of nation states, ranging from the Nazi obsession to complete an autobahn network to the unifying force of the 'Autostrada del Sole' that spines Italy, to the politicised nationalism of Thatcher's M25 (Sinclair 2002a, b). While I want to reserve discussion of such fixed infrastructures to the final chapter, it is nonetheless important to consider the idea that

> Motoring cultures may be distinguished by how they explore space and link spaces together via car driving . . . [which includes] stitching the nation together. The 'democracy' of car travel enables valorised national scenes and sites to be visited, opens up the possibilities for 'knowing' the nation . . . Previously remote places can be reclaimed within a national geography
>
> (Edensor 2002: 127)

Again, while automobiles have been accused of creating a generic landscape, it could be argued that these are generic *national* landscapes, as in the romance of American road cultures.

Fourth, there are a range of performances which revolve around driving that have a paradoxical effect on national ways of life:

> The independent mobility heralded by car culture has been partially desynchronising, disembedding familiar time-geographies from localities. However, the complex systems required for car travel have required a resynchronisation whereby the communal patterns of car use evolve: the routes used (for instance during rush hours and holidays), parking arrangements in cities, and common rituals. In concrete ways, shared cultures of automobility structure experience of time, and work, leisure and consumption patterns.
>
> (Edensor 2002: 129)

So, there is arguably a nationally-determined set of car cultures which perhaps enhance geographical distinctiveness due to the complexity of the driving process. This is linked to a stereotyped understanding of how other nations drive. As Michael (2001) has demonstrated, the emergence of the apparent psychosis known as 'road rage' condenses a diverse set of 'moral panics' in Britain. Interestingly, car and human operator merge as a hybrid agent known as the car driver. It is this agent that is undertaking – among other things – the claiming of national geographies.

Eating and drinking

> Like a language, food articulates notions of inclusion and exclusion, of national pride and xenophobia, on our tables and in our lunchboxes. The history of any nation's diet is the history of the nation itself, with food fashions, fads and fancies mapping episodes of colonialism and migration, trade and exploration, cultural exchange and boundary-marking. And yet here begins one of the fundamental contradictions of the food-nationalism equation: there is no essential *national* food; the food which we think of as characterising a particular place always tells stories of movement and mixing.
>
> (Bell and Valentine 1997: 168–9)

As with cars, as with football, as with everyday gestures and competences, eating and drinking are among the strongest elements of what constitutes a national community (Crang and Jackson 2001). Yet it is often regional traditions of cuisine that, after a slow process of diffusion in the national territory, become part of a national 'basket' (Bell and Valentine 1997). So, *paella* (originally Valencian/Mediterranean) becomes Spanish, as does Rioja (which is produced in a small region in the north of the country). As these products are gathered into nationally organised food systems, they become part of the modern household's dietary mainstays, even becoming associated with the exotic 'othering' of regional inhabitants (the Scots and Irish with their whisky, the Galicians and Bretons with their seafood) who may in fact be striving for secession or autonomy from a culturally overbearing nation.

Global media flows and national identities

> Driven now by the logic of profit and competition, the overriding objectives of the new media corporations is to get their product to the largest number of consumers. There is, then, an expansionist tendency at work, pushing ceaselessly towards the construction of enlarged audiovisual spaces and markets. The imperative is to break down the old boundaries and frontiers of national communities, which now present themselves as arbitrary and irrational obstacles to this reorganisation of business strategies. Audiovisual geographies are thus ... becoming detached from the symbolic spaces of national culture, and realigned on the basis of the more 'universal' principles of international consumer culture.
>
> (Morley and Robins 1995: 11)

In their book *Spaces of Identity*, Morley and Robins (1995) advance the argument that the increasing power of global media is provoking an 'identity crisis' within national media cultures. In particular, this involves a decline in public service broadcasting, which always had relatively strict quality controls, balanced programming, and as such 'became the central mechanism for constructing this collective life and culture of the nation' (p. 10). However, since the 1980s there has been a drastic restructuring of media industries, and with a wave of mergers and new developments in satellite broadcasting, the idea of a public culture has been undermined by a new set of commercial criteria. As such, these corporations push ceaselessly for a greater audience, seeking economies of scale in their programming.

The basic problem is that American corporate domination of the audiovisual sphere – especially in the emerging Multilateral Agreement on Investments (MAI) – plays on the idea of a 'lowest common denominator' in cultural terms, where advertising and the 'universal' stories of Hollywood are easily digestible by many cultural audiences (often inflected or indigenised through their own readings). The battle lines were drawn most notably in the GATT negotiations of 1993, the so-called 'Uruguay Round', which sought to remove trade restrictions and tariffs between nation states. As such:

> what we are seeing is the development of a postmodern battle between the US and Europe over images and sound. The US, employing its 'soft power' of commercialised culture, takes advantage of the forces of globalization by exploiting the global attractiveness of its language, its movies and music, its throwaway consumer ethics and its individualistic mentality ... Europe's definition of a rooted, 'deep culture' does not travel, does not resonate even across the board *within* its very own continent, let alone around the globe, so there is very little reason to assume that a European audiovisual sanctuary of some sort will be able to form a protective shield against ceaseless Disney-attacks.
>
> (Van Ham 2001: 83–4)

So despite the European Commission's attempts to create a European media space (Schlesinger 1997), the creation of a Euro-polity through media remains a long way away.

Distinctive national cultures are, it is argued, being attacked by a bland, intrusive capitalist modernity. Here, from the erosion of distinctive small shops to the domination of air-waves and television screens by global artists and film stars, to the impact of English on non-English languages, there is a sense of a threat to what makes a nation 'different'.

However, recent work has begun to demonstrate a complexity to this debate, in two main ways. In the field of pan-European advertising, Kelly-Holmes (2000) argues that while 'an increasing number of global advertising campaigns treat "Europe" as one un-differentiated segment, using pan-European media such as *Eurosport* and in-flight magazines to carry their messages' (p. 68), an awareness of the continent's linguistic diver-sity simultaneously allows for advertisers to play language games. For example, brands such as Ikea and Audi play upon the language of the country with which they are associ-ated to underline, respectively, the Swedish reputation for quality furniture design, and the technical quality of the German car manufacturing sector (captured in the '*Vorsprung durch Technik*' of their advertisements).

The second issue I take from Chalaby's (2002) discussion of pan-European television (PETV). While still subordinate to established national programming, the growth of satellite technology and market deregulation has seen the burgeoning of television stations such as MTV, Eurosport, Sky TV, Fox Kids, CNN, National Geographic, and BBC World since the 1980s. Such companies initially pursued a strategy based on economies of scale, assuming high levels of English language skills and the appeal of a trans-cultural programme content of sport and music. Squeezed by poor viewing figures and vigorous local challengers (successfully 'indigenising' their format), the pan-European channels began to adopt a range of strategies to diversify their output from the early 1990s: first, they began to introduce 'local advertising windows', allowing wider options for clients; second, the majority of PETV channels started to use dubbing or subtitling to appeal to diverse national language communities; third, local programming was introduced, to a limited extent, where the 'staple' output of the channel was supplemented by regional programming 'windows'; fourth, full 'local opt-out' possibilities exist, where a national channel is established under the umbrella of the generic 'parent', with its own news bulletins, purchasing of TV rights, and advertising (for example, French or British Eurosport). In the case of MTV, the establishment of competitor music channels such as VIVA (Germany) prompted a rethink of broadcasting strategy in 1994, with the launch of first three, then several more, separate European channels and national editions (Roe and de Meyer 2000).

Box 2   Euro Disney (Disneyland Paris)

While Americanisation has been a notable process in many European countries, it has been in France that the process has been most contested. Immediately after the Second World War, America and Americans were viewed positively in many countries of

Western Europe, especially for their role in the liberation of Europe from the Nazis. Yet this enthusiasm was not shared by all. While many young people embraced the new musical styles, fashions and lifestyle offered by the American film and media, others saw the American presence in more negative terms. In France, both Left and Right saw America as a threat – the Left, because of the introduction of assembly line working and fierce anti-trade union ideology, the Right, because of the perceived threat to traditional French values based on Catholic morality, the family, and – particularly – language (Kuisel 1993). As such, the American presence can be couched in terms of invasion, as follows:

- As a linguistic threat, where the strength of French as the international language of diplomacy was being swiftly undermined by the use of English (which in turn was bastardised in the French language).

- As an economic invasion: American multinationals were quick to seize upon the opportunities offered to reconstruct the shattered infrastructure of post-war Europe, and create new demands and cultural tastes among the populace.

- As a military threat: General de Gaulle jealously rebuilt the reputation of the French military after the Nazi invasion, and saw American-dominated NATO as a removal of French autonomy. Subsequent presidents have followed this policy, leading to the Pacific nuclear testing continued by Mitterrand and Chirac. The latter's refusal to join with the US in the 2003 Iraq invasion can be seen in this context.

- As media imperialism, the long history of French cinema and film-making was being undermined by the economies of scale offered to Hollywood, a tension arising in the GATT negotiations.

- As environmental degradation – the poor record of US corporations has been seized upon by French peasants, particularly in the defence against GM Foods.

These issues and prejudices were brought to a head in 1987, when the Disney Corporation signed a deal with the French government to build a theme park in Marne-la-Vallée, a new town on the eastern edges of the Paris city-region. For Disney, the operation was a logical extension of its operations beyond its American heartland. A hugely successful opening of a park in Tokyo encouraged it to implant itself in Europe. Yet the move was far more problematic than either side could ever have imagined. Three key issues emerged by the time of the park's opening in 1992.

(1) The power of the American (leisure) corporation

The French government, eager to stimulate its tourism sector and benefit from such a prestigious client, released several thousand acres of land to Disney below market value, as well as contributing to the costs of extending the regional and high speed rail lines between Paris and the park (Pells 1997: 306–9, Lainsbury 2000: 29–33). Disney was able to play off the French and Spanish governments – both covetous of the park as a magnet for regeneration – in order to secure the best financial deal. As a further source of tension, unions reacted strongly to the American working practices, dress codes, and Disney corporate mores, a long-running source of unrest between management and workforce.

(2) French attitudes to American culture

In Kuisel's (1993) discussion of the French relationship to American culture throughout the twentieth century, it is suggested that by the mid-1980s, the French had become enthusiastic consumers of American popular culture. The opening of Euro Disney split public intellectuals and politicians, between those such as culture minister Jack Lang who argued that Disney should not be seen in the same terms as William Faulkner or Jackson Pollock, and others such as André Glucksmann who suggested that Euro Disney was the 'return' to the 'old continent' of a culture spawned by European émigrés, from Charlie Chaplin to Fritz Lang (see Lainsbury 2000: 37–43). Yet fears remained that the intellectually dubious nature of the theme park model (with its selective approach to historical accuracy, its standardisation of popular culture, and deliberate social conservatism) was a motif for an increasingly 'dumbed down' Europe.

(3) An American narration of European history

> From the very beginning of negotiations with the Walt Disney Company, French officials
> insisted that visitors to the new park could not enter an entirely American world . . . That
> Euro Disney was stirring to life in the midst of Europe's efforts to reorganize itself into a
> single common market did not make matters easy . . . Recognizing that a dazzling display
> of Americana might be misconstrued . . . Jean-René Bernard, France's chief negotiator
> with the Walt Disney Company, made it his mission to ensure that Euro Disney
> respected 'European and French culture'.
>
> (Lainsbury 2000: 50–1)

It has to be borne in mind that Walt Disney's initial motivation for setting up a theme park was as a reaction to modernist urban development in the US. Its infamous Main Street USA – a sanitised version of American society as safe, conservative small-town life – was a clear statement of the park's American provenance. Yet the themed (or 'imagineered', to use Disney parlance) landscapes contained in the Paris park provided an interesting reflection of European (or French) encounters with American culture. For example, the 'Frontierland' (Western-themed) zone was seen as playing on a strong popular recognition of the cowboy stretching back to the beginnings of Hollywood. More complex was the retelling of European fairy tales to Europeans in 'Fantasyland'. The challenge to the designers was 'not simply to rebuild a Fantasyland that had worked in the United States or Japan, but to reinterpret it so that multinational audiences would not view the space as an Americanized bastardization of their own sacred stories' (Lainsbury 2000: 66–7). This highlighted the transnational borrowing of, for example, the Bavarian Neuchwanstein castle, and the difficulties involved in communicating to a trans-European audience in French (see Lainsbury 2000, Chapter 2).

Rebranding the nation

While many committed 'macro' nationalists have bemoaned the rise of the European Union as a threat to national cultures, it has been suggested by some commentators – most notably by Alan Millward (1992) in *The European Rescue of the Nation State* – that political elites in most European states have seen that the increasing complexity and integration of the global economy makes an idea of national sovereignty moribund. Nation-states, the argument goes, 'cannot hope to resolve such contemporary issues as nuclear proliferation, instabilities of world trade and finance, the power of multinational firms, international economic migrations and refugee flows, area conflicts and threats to the ecosystem. Only regional associations of states such as the European Union can attend to the increased scale of problems' (Hutchinson 2003: 37). And so by pooling sovereignty, they are able to exert some degree of control on other powerful actors (non-European nation-states and major corporations, for example). However, despite the apparent loss of political sovereignty, political parties, leaders, and their think-tanks and spin doctors are responding to this threat by repositioning their nation:

> Territorial entities such as cities, regions and countries are now also being branded like companies and products. The corporate brand has become an essential part of business identity, helping audiences identify with the company and – lest we forget – encouraging them to buy its products and services. In a similar way, territorial branding is seen as creating value in the relationship between territorial entities and individuals.
>
> (Van Ham 2002: 250)

Political branding has emerged as an important aspect of contemporary politics (Castells 1997a), with the election of the American presidency now won and lost on television. In Europe, there are signs that this approach is becoming dominant, and here I briefly discuss two examples of this branding process: Forza Italia and New Labour.

Forza Italia

Silvio Berlusconi's Forza Italia is probably the prime example of an explicit national rebranding. The 1993 collapse of the Italian political system due to widespread corruption left a political vacuum which was filled by Berlusconi, media magnate and president of AC Milan, one of Europe's most successful football clubs. Forza Italia – which won general elections in 1994 and 2001 – has been hailed as Europe's first 'postmodern' political party. Berlusconi's creation simultaneously repackaged Italy and offered policies based on market research and 'branded' much like any consumer product, in three main ways. First, it was geared almost entirely towards a strong media image – in much the

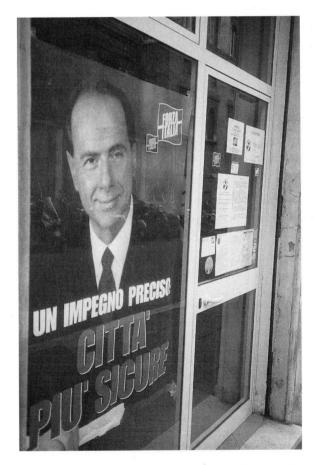

Figure 4 *Silvio Berlusconi: Forza Italia's election campaigns play on his personality to a huge degree. Donald McNeill.*

same way as the American presidency, Berlusconi and FI were presented as being coexistent. Second, the marketing of the party – its colours, the razzmatazz of its electoral campaigns, its name – is strongly related to sport, particularly football. The name Forza Italia roughly translates as 'Let's go Italy', drawn from football chants. Crucially, however, the Italian football team is seen as being one of the few unifying factors in Italy's 'banal' national identity. As the football team is largely mediated through its performance on television, so FI is able to align itself and promote itself on similar lines. Third, FI is promoted as a pragmatic, non-ideological party. However, with a strong business profile and a message based on affluence and material abundance, rather than on public services and redistribution, the party is clearly located within the political right. Ginsborg (2001: 296) defines FI's programme as follows: 'neo-liberalism as its ideological warhorse, the freedom of the individual as the basic tenet for society, ever-increasing and publicity-driven family consumption as the crucial economic motor'. While apparently projecting a clean break from Christian Democracy, FI has inherited many of the old social bases of the party and while electorally successful has remained tarnished by Berlusconi's on-going attempts to use prime ministerial privilege to curb the judiciary's investigations into corruption. The *Economist* famously characterised him before his 2001 victory as 'unfit to lead Italy' citing the conflicts of interest inherent in his huge and inter-locking media, property and football corporations, and his substantial implication in a series of legal battles (Ginsborg 2001: 318). However, by presenting this as in the clothing of a nationalist project of Italian economic revival, FI tries to distance itself from partisan party politics. (Koff and Koff 2000: 43–6).

Cool Britannia

The election of a Labour government in 1997 was seen as a radical shift in the power structure of the UK. With devolution of power to Scotland, Wales, and Northern Ireland, a more positive stance on European integration, and a relatively 'modern' approach to social issues, New Labour saw itself – spin doctors and all – at the forefront of a 'New Britain'. In the early years of the Labour government, strenuous attempts were made to associate politics with Britain's 'creative class' of pop stars like Oasis, 'Britart' artists such as Damien Hirst and Tracey Emin, and the high-earning sectors of design and, relatedly, tourism:

> So New Britain is 'creative Britain' in tune with the developments of a post-industrial information society. It has particular strengths in creative work and innovation, whether in science or in the cultural industries, industries crucial to a post-industrial economy dominated by services and the new information and communications technologies – an economy which New Labour aspires for New Britain
>
> (Driver and Martell 2001: 465)

Cool Britannia was never formally adopted by the government as a slogan, but it stuck in the media. And its significance is clear when one considers its message in contrast with the cultural image of Britain perpetuated by previous Conservative governments:

> Appeals to national identity are normally associated with Conservative politics: whether the 'putting the Great back into Britain' variety of Margaret Thatcher in the 1980s or the Baldwinesque images of village cricket and warm beer from John Major in the early 1990s ...To Labour modernisers these recent Conservative appeals to Britishness are seen as exclusive, nostalgic, and nationalistic, as well as being attached to traditional institutions, like the House of Lords, that have had their day.
>
> (Driver and Martell 2001: 461)

Yet New Labour's pronouncements were for many commentators an embarrassment (Bayley 1998; Rawnsley 2001). As I describe below, its use of an iconic building to aid this rebranding – the Millennium Dome – ended in disaster, and the attempt to 'rebrand Britain' has, in recent years, been toned down.

Urban landscapes and national identity

The early phases of nation-building in Europe, particularly in the nineteenth century, focused on a number of strategies. A common educational curriculum was important, clearly, as was the development of a nation-wide communications network aided by the railways and postal system. Military, economic, and legal power was also established throughout the nation's territory. Yet this idea of the building of a network to vertebrate the nation also usually relied on a clear process of central control not only of the immediate homeland, but also of empire. And as a means of achieving this, capital cities took on the role of government centre and cultural citadel. As Driver and Gilbert's (1999) collection *Imperial Cities* demonstrates, this era saw the rebuilding and planning of capital cities with broad boulevards, monuments, grandiose public buildings, and often accompanied by the staging of major commercial fairs and expos (Roche 2000).

In many ways, this centralised pattern prevailed through much of the twentieth century, not least in the post-1945 communist regimes in East and Central Europe. Here, cities like Bucharest, East Berlin, and Moscow were subject to bombastic building projects reflecting the power of dictatorial rule (see Leach 1999, for example, or Forest and Johnson 2002 on the Soviet monumental legacy in the post-Soviet Russia). Yet the new Europe shows every sign of continuing this interest in architectural debate, albeit in more democratic terms.

The contemporary European nation-state cannot be studied in isolation from its past. As Driver and Gilbert in *Imperial Cities* (1999) or Jacobs in *Edge of Empire* (1996) show, cities were central to the experience of European imperialism. Whether in Marseilles, London, Glasgow, Rome, Seville, Paris, or Lisbon (or, as we have seen, Brussels), the traces of the colonial relations of the nineteenth and twentieth centuries are clearly visible. In these works, we are introduced to the signature of material urban spaces of Europe's large cities – monuments, public buildings, urban planning, department stores both historically or in *contemporary* processes of urban restructuring (as with Jacobs 1994, 1996).

This is important, because many accounts of European identity have often been shy of discussing the urban grounding of the nation. As a brief example of what I mean, I want to consider the argument put forward by Kevin Robins (2001), who takes the case of London and Britain to argue that city and nation can be – perhaps – profoundly antagonistic spatial forms. In doing so Robins is careful to avoid a simple London versus the rest dualism, but argues that being urban and being national are two very different things.

> What is significant for me is that London has generally been left out of discussions of the national culture and identity – as if London were not properly, or purely enough, or manageably enough, British (or English, at that). And on those occasions when London has been referred to, then it has commonly been with feelings of resentment – resentment and hostility towards a city that seems to have a disproportionate share of national resources, and that dramatically overshadows the economic and cultural life of the rest of the country . . . And now, in times of global change and cultural disordering, it may be that the national resentment is deepening.
>
> (Robins 2001: 487)

Robins goes on to suggest that existing in cities is very different from existing in nations – the latter is 'a space of identification and identity, whilst the city is an experiential and existential space' (p. 489). In this sense, he argues, urban cultures are 'more provisional, more transitory and negotiable – less constraining and less sustained – than national ones' (p. 491). This latter statement is very contestable, but I wish to leave any discussion of that to one side for now. What is of immediate importance is the suggestion that cities are somehow more relevant – and are perhaps increasingly more relevant – than nations in contemporary Europe to the formation of identity.

Paris: Mitterrand's grands projets

Throughout the 1980s and the first half of the 1990s, France was presided over by François Mitterrand, whose controversial legacy was geared towards restoring France to a position of diplomatic and cultural centrality in Europe and the world. In a not dissimilar way to the imperial leaders of the past, he chose his own capital city of Paris as a cultural showcase. During his rule Mitterrand commissioned or saw completed around 10 major public buildings, lavishly funded by the state. These, ranging from the famous pyramid at the Louvre art gallery to the 'popular' opera at the Bastille to the French national library that now bears his name, were intended as reflections of a distinctive French identity that embraced bold, modern architecture (see Collard 1992; Looseley 1995 for a summary of many of these projects).

Mitterrand used these buildings as a means of 'theatricalising' Paris. As Kearns (1993) describes, the bicentenary of the French Revolution in 1989 saw a whole series of displays of French diplomacy and national pride, from military displays to the visit of the G7 world leaders (among 32 heads of state), who were treated to dinners and conferences staged in these new buildings. On the summit's first night, the Bastille opera was inaugurated, on the second, dinner was taken at the Musée d'Orsay. Finally, the summit continued at La Défense. By the time the celebrations peaked on 14th July with a grand parade of French military might beneath an Arc de Triomphe draped with a huge tricolour, the global media audience was huge: 80 TV companies, 700 million viewers, 3500 journalists. … with Mitterrand centre-stage. As such, Mitterrand and his *grands projets* provide a fascinating example of the remaking of national identity through apparently nineteenth century means. Indeed, it would not be too fanciful to compare his approach with the radical urban interventions undertaken by Hitler or Mussolini in Berlin and Rome, leading some to christen Mitterrand as enlightened despot, given his choice of bold architectural styles, his construction of a triumphal axis (as Hitler's architect, Speer, planned for Berlin) to La Défense, and his strong personal influence on the choice of architects in many cases.

Berlin, the Reichstag and German reunification

> Berlin is a haunted city. By the middle of this century, people living in Berlin could look back on a host of troubles: the last ruler of an ancient dynasty driven to abdication and exile by a lost war; a new republic that failed; a dictatorship that ruled by terror; and that terror unleashed on the rest of Europe, bringing retribution in the form of devastation, defeat, and division. Now that division, and the regime that ruled East Berlin, are also memories. But memories can be a potent force. … The calls for remembrance – and the calls for silence and forgetting – make all silence and all forgetting impossible, and they also make remembrance difficult.
>
> (Ladd 1997: 1)

In December 1990, Germany celebrated its first elections as a reunified nation, a little over a year since the opening of the Berlin Wall. After the euphoria, however, hard decisions remained. For 40 years, two different Germanies had grown apart under strikingly different economic, political and cultural systems. Overnight, they were thrown together again, with a resulting fear that the 'Wessies' would dominate the 'Ossies' in any number of spheres, from culture to employment to housing (Carr and Paul 1995). Geopolitically, the expanded German territory now had very important land borders with aspirant EU members, particularly Poland. And historically, questions were raised over the re-emergence of a powerful German state, a state often associated with Prussian militarism and the military-economic nexus that created Nazism. How was the German state to learn from its past, what was its collective memory? And how was it possible to reunify two very distinctive versions of national identity? Both within Germany and internationally, reunification posed huge questions about the nation's symbolic geography.

These fears and debates crystallised in debates over whether or not to move the capital of the reunified German state back from Bonn to Berlin, the divided city. In the aftermath of the Second World War, the Federal Republic of Germany (the West) was comprehensively decentralised, with its capital city located in the provincial town of

Figure 5 *Norman Foster's transparent dome in the Reichstag, Berlin. Donald McNeill.*

Bonn, as far from the bombastic military capital of Nazism as could be imagined, a 'city without a past' in the words of Konrad Adenauer, the first post-war German chancellor. Its humble rural location earned it considerable scorn from visiting journalists and dignitaries, and the provinciality of Bonn contrasted strongly with Berlin's world city status during the modern era. By contrast, the German Democratic Republic (the East) established its headquarters in East Berlin, where it had inherited many of the cultural institutions and grand architecture of the imperial capital. Through huge monuments and Stalinist building projects, the eastern section of the city was fashioned as a showcase for state communism (with West Berlin acting as a rather isolated neon showcase of Western German consumerism for much of the post-war period).

In 1991, after a long public debate which cut across traditional party lines, the German parliament voted by 338 votes to 320 to return to Berlin. In doing so, it brought to a head months of deep historical debate which surrounded the city's built environment.

What to do with the remains of dictatorial rule, as in the statues of communist leaders (Michalski 1998), the Palast der Republik (see Neill 1997 on this former state communist 'culture palace'), or surviving buildings from the Third Reich? How to commemorate or remember the murdered Jews of Berlin and Germany, and how appropriate is a single national war memorial as proposed at the Neue Wache (see Till 1999)? Should the city's 40 year divide be celebrated by the commercialisation of the Berlin Wall and the selling of prime land in the former no-man's land of Potsdamer Platz to Sony and Daimler Benz? Each of these questions has excited the attention of historians, politicians, intellectuals and, of course, Berliners since 1989 (see, for example, the debate between Marcuse 1998, 1999; Campbell 1999; Häussermann 1999; Ladd 1997; Richie 1999).

But in terms of the future German state, perhaps the most telling debate was over the shape of the future German parliament. As Wise (1998) has demonstrated, much of the argument was about the shift from Bonn as a 'capital of self-effacement' to a Berlin that aspired to world city status, historical maturity, and the needs of a complex contemporary capital city. Once it was decided that the parliament would be located in the Reichstag, the former parliament gutted by fire in 1933 (an event that prompted Hitler's assumption of total control of the political process) the debate about the appropriate *form* of a democratic architecture began in earnest. This was prefigured by the parliamentary debate over its 'wrapping' by the artist Christo (see Ladd 1997: 84–96). This, a two week event in 1995, was a simple (though technically complex) means of cloaking the building in a silvery aluminium-based cloth, rendering it a temporarily transformed artwork. By isolating the building's distinctive outline, and by creating a space of reflection and public fascination, the event both allowed the German and an international public to debate both the future form of the building, but perhaps as importantly served as a catharsis, allowing meditation over the building's role in Germany's troubled history.

The debate over the move into the old Reichstag building was equally contentious, given that the final bill for its refurbishment exceeded the cost of building a parliament entirely anew. The awarding of the contract to the British architect Norman Foster (one of 14 foreign architects invited to compete, along with numerous German firms) reflected the desire to internationalise such a potent symbol of German history. Responsible to a cross-party parliamentary committee, Foster was forced to constantly alter and modify his proposals, with concern surrounding – particularly – the skyline. For many, the nineteenth-century building's cupola was a symbol of Prussian imperialism, with Foster preferring to 'downsize' the parliament in relation to the wider public through the design of a public piazza above the chamber. The final compromise involved the construction of a cupola, but one that was transparent, and which allowed the public to view proceedings from above (Wise 1998: 121–34). By contrast with the the confrontational seating of the British House of Commons, for example, actual design features such as glass walls and circular debating chambers are seen as a means of downplaying political secrecy, reducing the physical status of demagogic orators, and emphasising collegiality.

While such an argument is perhaps simplistic, the new Reichstag has become a welcoming place for tourists and German citizens.

The Millennium Dome

Given the attention to branding epitomised by their 'Cool Britannia' strategy, it was little surprise that the 'New Labour' government of 1997 – that had sought to market itself as being a clean break with the Conservative governments of the 1980s and 1990s – chose to continue with the Conservative's aim of holding a major event to celebrate the millennium. It seemed to offer all sorts of advantages to a publicity-hungry government:

> If New Labour meant new hope; if this, as Blair had said in his party conference speech, was a 'young country', if the river Thames could be seen as a string along which the regenerative pearls of Richard Rogers' vision of the new London could be strung; if the belt of poverty in east London was to be broken; if one wanted to believe in the future; if one wanted to say something about Britain not only to British people but to a wider audience, presenting the country with a different perspective to the world; then what better vehicle could you look for than the Dome?
>
> (Nicolson 1999: 105)

As such, on New Year's Eve, 1999, the Dome was opened to a VIP audience, including the Queen, leading political figures, and major players in the arts and media, including the editors of all major newspapers. So, as an attempt to 'rebrand' Britain the government was clear in its attempt to maximise the impact of its new creation. There are four points to consider here:

- The choice of building – a Dome designed by Richard Rogers Partnership – was bold and hi-tech, seemingly reflecting the desire for New Labour to embrace a similar future for Britain.

- The choice of location – London selected over Birmingham – reflected the desire to restate London's 'capitality' of the New Britain, perhaps subconsciously reflecting the sustained centralism of Labour policy-making, and indicating the importance of the capital city at a time of political devolution.

- The choice of content – heralded as a celebration of British culture – was the key issue, and one that Labour hoped would echo the popularity of the earlier national jamborees.

- The huge investment of public money made in the project. Many in Britain had sympathy with marking the millennium, but not with the wasteful way the project was run.

Following a disastrous opening night, the Dome was berated in the national press. Visitor numbers were far below that predicted by the organisers, meaning that the cost to the exchequer was never recouped. Yet this was despite the fact that in their anxiety not to leave the taxpayer with an excessive bill, whole sections of the Dome were splashed with adverts and sponsorship of major firms, from Boots to McDonalds to Ford.

This was all mercilessly satirised by critics. Andrew Rawnsley summarised it thus:

> Not in the way that was ever intended, the Dome did symbolise Britain at the turn of the century. The Great Exhibition of 1851 was a celebration of the industrial and trading prowess of mid-Victorian Britain. The 1951 Festival of Britain was an uplifting tonic after

Figure 6 *The Millennium Dome – desolate hi-tech. Donald McNeill.*

> wartime adversity and austerity. The Greenwich blancmange also distilled the spirit of its
> era, but the least attractive aspects of the time. It embodied the most meretricious
> features of the consumer age which New Labour had absorbed too well. The Dome was
> the vapid glorification of marketing over content, fashion over creation, ephemera over
> achievement.
>
> (Rawnsley 2001: 327)

The reason for such bile was clear. Stephen Bayley – who resigned from creative directorship at the Dome – mischievously noted that the Labour ethos of policy-making by market research was incompatible with the greatest artistic achievements: 'One imagined that a focus group would have turned down the Great Pyramid (impractical), Beethoven's late quartets (too difficult) and [Picasso's] Guernica (too inflammatory)' (Bayley 1998: 139). Few took satisfaction from the disaster, even those who had always pointed to the potential for the building to become a white elephant. Yet the whole sorry affair indicated the extent to which landscape symbolism and the big event – the 'national party' – remain central to the rebranding of nationalism in contemporary Europe.

Allan Pred and the globe/alisation of Sweden

> The Globe, in particular, lends itself to metaphoric and metonymic images, being a
> crystallisation of – and contributor to – the ongoing processes through which Sweden's
> commodity society is becoming further globalized and further subject to unchecked
> market rule ... [or] ... the identity 'crises' which have pervaded Sweden during the late
> 1980s and 1990s. Time and time again, the images – and related discourses – evoked by
> the Globe directly or indirectly link up with a pair of fundamental questions:
>
> > Where in the world am I, are we?
> > Who in the world am I, are we?
>
> (Pred 1995: 202)

In *Recognising European Modernities*, Allan Pred (1995) provides a notable example of experimental urban writing. By invoking Walter Benjamin and his use of a montage of quotes, critical commentary, and – as excerpted above – home-grown poetry, Pred provides a fascinating, if at times exasperating, journey through modern Swedish national identity as played out in Stockholm. In his study of three contrasting consumption spaces from differing points in the preceding 100 years (the Stockholm exhibitions of 1897 and 1930, and the Globe sports and shopping arena of the 1990s), he charts the rise of advertising and the commodity, linking this to notions of a stable, social democratic Swedish political model. As in his later book, *Even in Sweden* (Pred 2000), which is an account of racism in Swedish society, his narrative of the globe is designed to demon-strate the unsettling nature of globalised economics and cultural and human flows on one of the most cohesive national societies in Europe.

The Acropolis

When one searches for the archetypal monument that symbolises national identity, the Acropolis in Athens is difficult to beat. While trying to understand monuments through a reading of their formal aesthetic properties is usually a fruitless task, through ethnographic method and social history the significance of these sites can be revealed (Jacobs 1994, 1996 is an exemplar of this method). As with the Reichstag or the Dome, the political and cultural battles over the Acropolis through the twentieth century reveal a fascinating insight into modern Greek national identity. Yalouri's (2001) discussion places the Acropolis in the context of its curious historical nature, embedded – as in the case of Rome – within a 'local present' but with a 'global past' (p. 5). Here, the Acropolis is a 'condensation site of national identity, and a meeting point of classical and contemporary, diasporic and mainland Hellenisms' (p. 25). Yalouri documents the various media and political controversies generated by, for example, the replacement of the Parthenon columns by Coca-Cola bottles in an early 1990s advert, the claims of 'repatriation' of the Parthenon (Elgin) Marbles (held by Britain), or the rights of the modern Greek state to claim the patrimony of the ancient past.

Figure 7 *Acropolis © Audiovisual Library European Commission.*

Europeanising the nation

It is important not to be over-enthusiastic about the idea of the simple 'death' of the nation, therefore. It makes more sense to see nation-state forms as becoming more fluid, shape-shifting, and strategic, as political parties reposition themselves in terms of a new global order. We have looked at the examples of the repackaging of notions of Italy and Britain in recent years. But there is another, not unrelated phenomenon being played out in, particularly, the once-peripheral nations of the New Europe. This is the direct embrace of European integration not as a source of rivalry or fear, but rather as an opportunity to escape from archaic and conservative notions of national identity. This is certainly marked in many of the countries of the post-communist world. But it has perhaps been most clearly marked in two countries which have actively seized European 'club' membership for cultural, as much as economic, reasons, and where 'Euroscepticism' has been almost entirely absent from political debate. These two cases are Ireland and Spain.

Ireland: between America and Europe

Since 1973, when Ireland joined the European Community, enormous changes have been afoot in this once predominantly agricultural nation-state. It was then famous for the huge levels of out-migration by many of its young people, seeking job opportunities and greater cultural freedoms elsewhere. While Britain, and particularly London, was often the destination of those disaffected by the stasis of the Irish political class, a predominant focus of cultural identity was the United States, especially its eastern seaboard and the huge Irish ethnic communities of New York and Boston. By the late 1990s, the situation had changed. While the English-speaking neighbours to its east and west were still important, culturally and economically, Europe was becoming of increased significance. There were at least three reasons for this:

- First, the rejection of the prevailing sense of Ireland as a deeply Catholic, agricultural nation that looked back to its essential cultural identity, and its replacement with a secular, technologically progressive state and economy and sense of a renewed political identity. This culminated on the one hand in the election of Mary Robinson as Ireland's first female president, and on the other in the approval of the Maastricht Treaty where Ireland's ability to opt out of European abortion legislation became the dominant topic of debate.

- Second, the overwhelming importance of EU funding through the Structural Funds, that saw the modernisation of Irish infrastructure and agriculture. The rejection of the Nice Treaty by the Irish electorate in summer 2001, seen by many as an anti-immigrant, anti-enlargement vote – a 'drawbridge' vote – reflected the extent to which the Irish electorate sensed that EU membership had delivered them considerable privileges.

- Third, a remapping of Ireland as being European, rather than (or as well as) American in terms of its external focus. As O'Toole's (1994) thoughtful discussion of the new map of Ireland shows, the way in which nationalist ideology is constructed depends upon a very real drawing of Ireland's place in the world. So, while the north and south of the country are famously divided, so the contrast between east (Dublin-dominated) and west is also of utmost importance:

> Strongly reinforced by the intellectual elite of early twentieth-century Ireland, the 'West' became an idealised landscape, populated by an idealised people who invoked the representative, exclusive essence of the nation through their Otherness from Britain.
>
> (Graham 1997)

This – 'Irish-Ireland' – had by the 1960s begun to disappear in the rest of the south, and had no correlation with the industrialised, Protestant north-east counties. Similarly, a split has long been evident between the largely agricultural Irish-speaking west, in areas such as Galway, and the urbanised east of Dublin and its environs (Nash 1993).

Thus the nature of 'Europeanisation' has come to influence much contemporary debate on the future path of the Irish nation, and such opening to outside forces is seen as being both an escape from the constrictions of Rome-led religious orthodoxy (on issues such as birth control, divorce, and gay rights, for example) and a source of economic regeneration. Furthermore, as MacDonald (2000) has discussed, this includes very real decisions over the nature of urban living, and whether Dublin as a capital city can cope with its phenomenal spurt in return migration, bringing urban sprawl and huge problems of traffic congestion. The place of Dublin as a cosmopolitan city within Irish culture has been elaborated by, for example, Kiberd (2002), Lentin (2002), and Maguire and Hollywood (2002). For Lentin (2002), Dublin is now in a complex moment of its history, a time in which the city has shifted from being a capital of a nation of emigrants, to being the centre of a diaspora of migrants from Romania, from Nigeria, and elsewhere. In short, as Ardagh (1995) describes, the Europeanisation of Ireland has permeated debates about the country's 'traditional' local identities, be it linguistic, social or economic, seen most clearly in the way Ireland and its landscape are represented to tourists.

As in many cases, these tourist representations or 'gazes' (Urry 1990) use a range of clichés that draw upon an essentialised notion of the national society. For example, Markwick's (2001) study of the images used in Irish tourist brochures lists a range of images which can be summarised as follows:

> photographs of distant green hills and blue-tinged mountains provide a background to a lough or other water body; a patchwork of green fields and gently rolling hills, sometimes dotted with small farms; an ancient building or traditionally white washed and thatched cottage; a traditional village pub . . . Local inhabitants, often elderly men in tweed caps, appear to be shown engaging tourists in conversation . . . ; a deserted coastal scene; an 'activities' picture . . . most often of golfers on emerald green coastal links . . . ; an image of a cityscape (either Dublin or Cork) focusing on buildings of historical and architectural interest; other views that appeared less frequently included stereotypical shots of young irish dancers or farmworkers, farm animals, typical Irish signs, and foods or goods.
>
> (Markwick 2001: 41)

From this summary, Markwick argues that we are witnessing a 'commodification of Ireland', based around three main principles:

- First, images used in tourist literature tend to be of sparsely populated areas. It is ironic, Markwick notes, that the poverty and underdevelopment from which many Irish have had to move – whether to Britain, the US, or elsewhere – has left behind an 'empty space', areas open for tourist escape.

Second, the images of Irish people in many of these representations are highly idealised, with locals presented as 'smiling, friendly, and consequently "hospitable"' (Markwick 2001: 42). Employment activity is often presented as being close to nature, with traditional agricultural tasks such as turf-cutting being presented as the dominant economic activity. As such, these representations 'play on themes of difference, drawing on oppositional constructs of work and leisure, urban and rural, and "core" and "periphery". Ireland and Irish people are represented as "other" to the workaday world of the tourists' who tend to visit from the metropolitan core. As such, the presentation of Irish society is very partial, perhaps playing upon a patronising view of the inhabitants.

- Third, Ireland is represented as 'timeless, unchanging and firmly embedded in the past' (Markwick 2001: 42), communicated through the prevalent use of landscape images and Celtic motifs.

However, while this may be the predictable response of tourist marketing agencies, there is another side to representations of Ireland that advertisers have seized upon which understands that many audiences have a sophisticated, ironic understanding of such clichés (Markwick 2001:46). Various television advertising campaigns have satirised, for example, the Riverdance stage show of Irish traditional dance, or old men in tweed caps. Markwick (2001) concludes by suggesting that there is a ready market – both in the Irish diaspora and among non-Irish – for believing in a mythic, romanticised Ireland. It could be argued that as Ireland becomes increasingly developed – old stone cottages replaced by modern bungalows, for example, are a ready feature of the 'timeless' west coast – so a nostalgic vision of the past becomes increasingly popular.

Post-Francoist Spain and 1992

While Ireland had been kept on the periphery of Europe due to a sluggish economy and a culture of out-migration, the Spanish case – as with those of Greece and Portugal – is rather different. Until the mid-1970s all three of these southern European countries had suffered totalitarian dictatorship, Spain being the most famous. Between 1939 and 1975, the country was ruled by the victorious Nationalist grouping of the Civil War, a coalition of royalist, capitalist, military and Catholic social groupings that resented the Republican working class movements (which often overlapped and allied with the separatist movements of Catalonia and the Basque Country). The Francoist regime had been unable to modernise Spain sufficiently within a changing European context, which had left the country culturally isolated and economically stretched. On both sides of the ideological divide, many of the country's major employers argued for European membership to exploit other European markets, and many on the Left argued in parallel for the secular and modernising opportunities offered by European integration. As Jauregui (1999) notes in a historical comparison between British and Spanish attitudes to Europe, 'the dominant paradigm of *Spanish* national greatness has incorporated the concept of "belonging to Europe" with much greater ease than in Britain (p. 281)'. Here, membership of the EU involves a fundamental remapping of Spain where the borders – once defended by Franco to the extent of having a separate rail gauge – soon became a fundamental part of Spanish culture (it should be noted that Catalan and Basque groups had their own distinctive reading of their place in Europe).

This eased a relatively smooth transition to democracy, with conservatives appeased by the restoration of a parliamentary monarchy under King Juan Carlos I, and the re-establishment of a multi-party political system. Under the much-admired autonomisation process, many of the political claims of, particularly, the Catalans and Basques were satisfied. By acceding to European Community membership in 1986 (along with Portugal and shortly after Greece) the position of Spain within Europe was dramatically transformed. The economy boomed in the late 1980s, and while suffering from global recession in the early 1990s has generally recovered to become as stable as the European norm.

This gave rise to talk of a 'New Spain', one of material affluence and dynamic cultural production. There have been claims that the nation has returned to the situation of 1936, where on the eve of the Civil War Spain was a 'cultural superpower', giving the world some of its greatest modern painters – Picasso, Miró, and Dalí – along with one of its finest poets, Lorca. The return of democracy – and generous subsidy to art from central and regional governments – has meant that Spain once again became very fashionable. This began with the events of 1992 – the Barcelona Olympics, Madrid's spell as European City of Culture, and the Seville World Expo – which saw the radical transformation of the urban fabric of, particularly, the Catalan and Andalusian capitals. Yet these events, so global or European in their outlook, perhaps masked a deeper cultural tension. Their intention was straightforward: the 1992 events ...

> were explicitly intended to celebrate Spain's coming of age as a modern, democratic European nation-state, marking the end of a period of political transition (and uncertainty) ... But these popular celebrations of Spain's new status tended to neglect the past and glorify the present. Indeed, this seemed to be part of an official attempt to represent Spain's new, 'modern', democratic national identity as if it were built on a *tabula rasa*, thus avoiding confrontation with the cultural, social, regional and political tensions that have plagued Spain since its emergence as a nation-state.
>
> (Graham and Sánchez, 1995: 406)

It has been suggested that Spain has been unable to come to terms with its past, that it has suffered a 'collective amnesia' where the deeply fractured social, regional, and ethnic faultlines that were played out in the Civil War have been forgotten for the sake of political and social stability in the present.

So what is the significance of Europe for the Spanish nation? As with Italy, the 'normalisation' of the political system has been enhanced by Euro membership: for example, 'the implications of corruption have become inescapable: the greater the corruption, the higher the budget deficit; the higher the budget deficit, the less the government's scope for generating employment and prosperity, whether by cutting taxes or boosting investment' (Hooper 1995: 440–1). Culturally, there is a sense that the languages repressed under Franco's centralist regime can flourish again. In the field of international and European affairs, Spanish politicians and lawyers have played a significant role: Javier Solana became president of NATO, and Chilean ex-dictator Augusto Pinochet's extradition saga was led by Spanish prosecutor Garzón. There is a sense, therefore, that the New Spain has mirrored the emergence of a New Europe, one in which old traditions and conflicts are being wiped away by a tabula rasa of economic modernisation, and technocratic inter-governmental policy-making.

Conclusion

It is often argued that European integration is eroding the nation-state. As Hutchinson (2003) suggests, this is misleading for three reasons:

> First, national identities often predate the era of the modern state, and the persistence or intensity of national identities cannot be explained by the success of state-led modernisation, because the modern period is also one of disruption to state authority. Second, the pooling of sovereignty is not a revolutionary new development since nations have continually varied in strength and in the degree to which they wish to regulate the sectors of social life. Third, many, if not most, European national identities have been developed either alongside or in relation to a sense of Europeanness, and most conceptions of 'Europe' arise out of prior national views of the world.
>
> (Hutchinson 2003: 37)

Bearing this in mind, this chapter pursued the idea that European integration is enhancing, not dissipating, the symbolic identity of the nation-state. However, this is taking place in what could be called a postmodern context, which reduces the 'boundary consciousness' of many individuals. Here, centralising nationalism (in the spheres of language, education, defence, taxation, public sector broadcasting, and so on) which accompanied the classic formation of the nation-state is undermined by a variety of globalising tendencies. However, national political elites may be using these flows to represent the nation, as in the rebranding processes identified by Van Ham (2002), but also in the wave of redesigns of national parliament buildings – headed by the debate over the new Reichstag and the shifting of the German capital from Bonn to Berlin – or in Spain and in Ireland, where the mark of European identity has transformed archaic, 'primordial' notions of nationhood with new discourses of modernity, a knowledge economy, and innovative cultural production.

Various destabilising tendencies have impacted significantly on the nation-state as we know it. While nations seek to promote an ideology of timelessness, they are also forced to face challenges to long-held ideas of sovereignty, from currency to language to borders. In a period where European integration is increasingly intense, with substantial migration flows, transnational identities, with cross-border media, new technology and economic loyalties, national identity has been turned upside down. Yet while this might signal *decline,* some national governments are actively seeking to remake their identities. In an extreme case, this may through a politicised process of branding, as in Berlusconi's Forza Italia. However, almost all nation-states are looking to their major cities, and especially their capitals, as a showcase for their new direction. Nonetheless, national identity is being altered, and its connotative landscapes are now – as during the imperial era – often displayed in the restructuring of major cities or in infrastructural projects.

Finally, the 'map' of the nation is no longer the easily defined schoolbook version with clear borders. The physical world of mountains and rivers is transformed by pollution and tourism. The economic world – never easily contained within national maps – is now almost unmappable, given the speed and virtuality of currency flows and investment decisions. Cities, regions, motorway, and rail and air corridors and their ports and interchanges, borders and borderlands, all of these are transforming the national story.

3
Regional renaissance

One of the most prevalent clichés in European politics in recent years has been the notion of a Europe of the Regions. The Euro-federal model argues that along with an overarching European government led from Brussels, there is scope for devolution of the nation-state's powers to a lower level, somehow 'closer to the people'. It also has a strong economic rationale. Recent research has argued that regional economies are 'leaner', more sensitive to fluctuations in market demand, fitting with a post-Fordist production paradigm of skilled labour producing high technology products. Culturally, it means that 'historic nationalities' with centuries-old traditions of speaking a language or dialect, a distinctive cuisine, or traditions of self-government can regain that identity that has been eroded by nation-state education, economic and cultural policy. Politically, few would argue that bringing democratic control to a smaller level of population size is a 'bad thing' in terms of encouraging a participative democracy.

However, many commentators have detected a darker side to this image (e.g. Fraser 2000; Morley 2000). Several of the new regionalist parties such as the Lega Nord in Italy seek to create a buffer against globalisation – and this can mean xenophobic, anti-immigrant policies. They may, similarly, be hostile to the classic Keynesian/social democratic notion that the wealth of one region can be shared or redistributed to the poorer societies on the national or European periphery (Hadjimichalis and Sadler 1995), what Harvie (1994) calls 'bourgeois regionalism', suggesting a smug vision emanating from Europe's prosperous core. The fragmentation of the post-communist states has seen ethnic cleansing and civil war in Yugoslavia, and the rediscovery of potentially chauvinistic national traditions in the likes of Slovakia and Belarus.

In truth, the story is more complicated than either of these two positions. The impact of globalising trends on Europe – whether in the Americanisation discussed in the previous two chapters, or the increasing tendency of corporations to seek direct negotiation with regional governments – is not a one-way flow of power. Regional governments are both embracing and rejecting these trends, sometimes simultaneously! And the nation-state has not disappeared – it still, in almost each case maintains a very powerful role in determining the fortunes and scope for action of the regions. In the post-communist states, the nation-state is enjoying a renewed lease of life. And yet, particularly in the affluent states of Western Europe, there is either a well-established set of regions with substantial autonomy (Germany) and identity (Spain), or else with a growing sense of collective pride and political possibility (UK, France, Italy).

This chapter is structured as follows. First, I provide a typology of differing kinds of regional identity, again highlighting the diversity that exists within European regions. Second, I highlight the processes that might be strengthening the power of regions: the 'democratic deficit', the desire to preserve cultural and linguistic identity, and the rationale behind regional economies. Third, I discuss several case studies of some of the most dynamic aspects of contemporary European regionalism, focusing especially on areas in which there is a clear engagement with the 'global'. I profile the establishment of the Bilbao Guggenheim art gallery in the Basque country, whose nationalists have been particularly adept at relocating their territory within global art and media networks. I then question whether regionalism is a politically progressive phenomenon (in the sense of being inclusive, democratic, and cosmopolitan), looking particularly at the 'invention' of the region of Padania by the Italian Lega Nord, which fuses economic prosperity and hostility to 'other' regions in the Italian South. This is followed by a brief outline of the impact of Braveheart, the Hollywood film, on Scottish identity. Finally, I explore how the competing versions of Catalan identity which are existent between a cosmopolitan, Barcelona-based conception of modern Catalanism, and a more ruralist, conservative script were played out in the staging of the 1992 Olympics. This final case study is crucial in showing how cities may actually compete with regions, rather than there being a general, unproblematic process of 'locals' against the nation-state.

A Europe of the regions?

When we begin to shift our eyes away from what Chris Harvie (1994) calls 'school history', the taking for granted that the nation-state is the primary and even singular form of identity, we see the European map morph dramatically. No longer do we have the homogeneous nation-state with its partner people, the jolly Germans, romantic Italians, promiscuous Swedes (Enzensberger 1989) but rather a more complicated, analytical picture. On the face of it, it is quite easy to break down. Most contemporary nation-states have either federal structures (particularly Germany and Belgium, and Spain in a less comprehensive way) with clear powers of autonomy for the regions, or else are moving towards an administrative decentralisation which nonetheless retains significant powers in the centre. Yet when one begins to look at the relative wealth, territory, and population sizes of Europe's regions, to say nothing of the strength of their identities, a very complicated patchwork emerges. Furthermore, due to the processes described in the previous section these are *dynamic* – it is not a matter of whipping off the national blanket developed over the past couple of centuries to unveil this multi-coloured quilt beneath, but rather to remember that this Europe of the Regions is in the process of being *made* and *un*made.

Rather than fully document the diversity, I want to highlight various trends here, of which I note five:

(1) The *historic nationalities* – in Spain, Catalonia, the Basque country, and Galicia; in Britain, Scotland, and Wales; in France, Brittany, and Corsica; in Belgium, Flanders, and

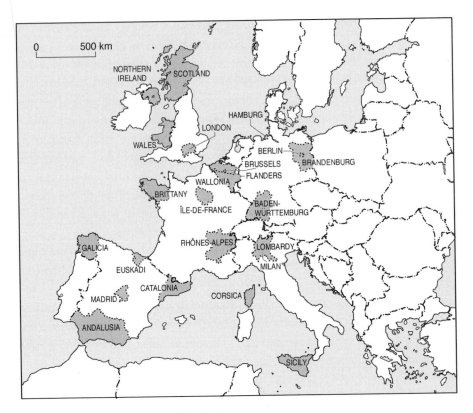

Figure 8 *Key spaces in a Europe of the Regions.*

Wallonia, in Germany, Bavaria. Here, for reasons of linguistic vitality, traditions of self-government, warfare, a shared sense of hostility to central government funding decisions, or sensation of underdevelopment relative to core regions, are places which have centuries of existence as polities in some form or other. Fundamentally, these historic nationalities have electorally popular political parties that lobby for more autonomy or, indeed, independence from the nation. This is a very important point: most strands of nationalism have moderate and extreme versions. The Basque country, particularly, possesses the separatist extremists of ETA and its political allies, but the most popular party is the moderate, yet still pro-separation Partido Nacional Vasco (PNV). There may be as much animosity between such groups as between centralist parties and the periphery.

(2) The *city regions or city-states*: such as the Comunidad de Madrid, Île de France (Paris), Brussels, the Länder of Berlin, Bremen, Hamburg, the Greater London Authority. These conurbations of millions of people often dominate the economic and cultural lives of their nation-states, and hence have a strong voice in any policy and resource debates and lobbying activities. Furthermore they are likely to be very ethnically diverse. By contrast with the argument that regionalism is awakening hostility to outsiders, these city regions may in fact be pioneering new forms of social life,

generating hybrid identities, possessing different speeds and forms of living from other regions within the nation-state. Being essentially *urban* in nature, they are a fundamental part of the Europe of the Regions/Europe of the Cities debate, but are poised in-between the two. Yet an understanding of their geopolitical and cultural significance is often subordinated to research into their economic role. I briefly discuss how these tensions might be manifested in the Barcelona vs. Catalonia debate at the end of the chapter.

(3) The *administrative inventions*: Rhineland-Palatinate, Baden-Wurttemburg, Rhônes-Alpes. These are particularly interesting, in that many French and German regions – some of which have become very successful – have very little historic identity as cultural or political 'units'.

Fourth, we can speak of the *micro-states* such as Andorra, Channel Islands, Monaco, Liechtenstein, San Marino, Isle of Man, the Vatican. And fifth, the *islands* of Europe, such as the Balearics and Canaries (Spain), Sardinia and Sicily (Italy), Corsica and Martinique (France), Madeira, Cabo Verde, and the Azores (Portugal), and so on.

However, across these groups there is no uniformity, so there are two caveats to this discussion. First, a Europe of the Regions is very diverse (*New Statesman and Society* 1992). For example:

- Portugal has a population of around 10 million. Île-de-France has around 10 and a half, yet Portugal gets full membership of the European Union, while metropolitan Parisians have to take their chances with the French state (which isn't always too difficult, admittedly).

- The Vatican City has a population of less than a thousand, yet runs the world's largest institution (the Catholic Church) in terms of active participants, and is probably the world's single most visited tourist destination.

- The likes of Catalonia and Scotland have traditions of democratic government that predate those of nation-states by centuries, yet have seen such rights eroded by successive centralising nation-states.

- The territory of Andalusia (population 7 million) is 87,268 sq. km, making it larger than Denmark and the Netherlands amalgamated.

- Corsica is 204 km from Nice, but only 109 from Livorno in Italy, yet is governed by France.

- The Brussels city-region hosts the European Union's main functions, especially the Commission, as well as NATO. It is also the regional capital of Flanders, and the Belgian nation-state.

I only flag up these examples as a means of illustrating both the difficulty in talking of a Europe of the Regions, and to suggest why – historically – such a curious patchwork exists.

The second thing to bear in mind is that of political power and importance. Charming though the Azores may be, it is clear that there is an elite group of regional actors that house, appropriately enough, many of the elites of the New Europe. For example, the 'Four Motors', comprising the regional governments and development agencies of Lombardy, Baden-Württemberg, Rhônes-Alpes, and Catalonia, raised eyebrows by their formation of a co-operative alliance aimed at advancing their economic potency. While these regions are typically centred around key 'second cities' of Italy, Germany, France, and Spain, they jostle for resources with the city-regions that dominate spending decisions in their respective nation-states, Rome (and the South of Italy), Berlin-Brandenburg, Île-de-France based on Paris, and the Comunidad de Madrid. It is partly what happens here that will determine the potency of any regional renaissance in Europe.

Why the resurgence of regional identities?

So what accounts for the enthusiasm for this Europe of the Regions? I think we can break the debate down into five main areas

The democratic deficit

We saw in chapter 1 that the emergence of a remote, bureaucratic state in Brussels that is taking over national functions 'by stealth' has given rise to the genuine fear of a 'democratic deficit' that is not filled by the European Parliament. Decision-making – made in tortuous fashion by the nationally constituted Council of Ministers – is thus indirect, and it is not clear from national political programmes what kind of impact a European directive or policy will have. As a response, the idea of 'subsidiarity' was introduced in the Maastricht Treaty – that policy decisions should be taken at the appropriate level, and not always from above. This gave support to the arguments of regional movements that they were not leading their citizens away from the protection of the nation state to a situation of isolation or autarky, but that the European Union could provide an umbrella that allowed them to pursue, for example, educational, linguistic or cultural autonomy while ensuring economic prosperity. This belief was clearly demonstrated in the Scottish National Party's 'Independence in Europe' slogan of the early 1990s. Consequently, pressure from regional and urban entities across Europe, but particularly in Germany, led to the creation of a new body of the EU, the Committee of the Regions. While unelected, and restricted to providing opinions on how EU legislation would feed down into sub-national bodies, it has been seen by some as the first step towards the federalist model of European governance.

Preservation of cultural identity

In this sense, then, regions could reinstate their own, indigenous traditions – especially of linguistic broadcasting and political self-determination – while not being tied into the economics of the nation-state. However, regionalism has also been associated with a return to 'primordial' nationalism, the erection of a defensive community drawing tight borders between 'them' and 'us' (Evans 1996). While this might have liberatory overtones (radical Basque separatism is premised on a clearly Marxian anti-imperialist ideology), it too often descends into xenophobia. As Fraser (2000) has suggested, this is the Europe of the 'lost tribes', the idea that linguistic communities, or areas with clearly distinct histories from their national-states, or the deeply-regionalised identities that form part of successful national parties (such as the strands of British Conservatism so cruelly exposed by the Maastricht Treaty), are coming under attack by EU standardisation and more general processes of globalisation. In an uncertain world, plain-speaking leaders can animate the politically fuddled of Flanders and Varese. As I shall discuss below, many of these parties or movements have rather dubious political agendas: the Italian Northern League led by Umberto Bossi has a distinctly racist, sexist agenda. Extreme Flemish nationalism is based on an anti-immigrant platform. But it also must be borne in mind that at a time in which nation-states are no longer delivering the goods, devolving power to the regions or historic nationalities may also be a way of improving democratic com-munication, or even of slipping neatly into a European federalism. So while some of these defensive groups may be anti-Europe, some believe that actively feeding into the federalist project might be the best way to preserve their fragile cultures.

This strong sense of place identity has been identified with a political vocabulary that emphasises cultural roots and purity, and the importance of borders, a situation that was exemplified in the ethnic cleansing of the Balkan wars. There is no doubt that commentators see this in terms of a *revival,* that as the world is increasingly globalised, standardised, and political representation appears to slip ever further away, so have come a wave of movements that seek comfort in a strong sense of identifiable place. I have already spoken about the 'othering' that this might entail, and the likes of Morley (2000) have drawn attention to the German concept of '*Heimat*', or home, with connotations of belonging or security. Very importantly, too, 'the over-valuation of home and roots has as its necessary correlative the suspicion of mobility' (Morley 2000: 33).

As I discuss below, then, this 'suspicion of mobility' may be used in contradictory ways. Many of the most dynamic regions combine economic strength with attempts to keep down taxation and social spending, thus rejecting immigration from other parts of the EU or beyond. Or they may combine cultural chauvinism with a sophisticated understanding of global markets.

Linguistic identity

The penetration of English into the European Union is, of course, only a small part of the general impact of English in countries around the world, be it Latin America, Africa or Asia (see Crystal 1997). Here, however, I focus specifically on the impact that English has on the European Union. First, the 'big languages' noted by Graddol (1996) are in tense competition: English, as the language of one of the most Euro-sceptical nation-states in Europe, is thus seen as inappropriate for EU business; French tends to predominate as the working language in EU institutions. However, German – as the language of the most economically dominant state – is also important, particularly in informal business communication in Northern Europe, and will be strengthened by the accession of Austria and future enlargement to Central and Eastern Europe. Thus while the 12 national languages are officially seen as having equal status, there is a clear hierarchy involved.

What is clear, however, is that in recent years – and with the encouragement of sections of the EU – there has been a renaissance of minority languages with official state recognition (for example, with state-licensed education in that language), and of those seeking rights for dialects (such as Lallans, the dialect of Scots English) and of sizeable ethnic minorities within nation-states that speak other languages, such as the Turks in Germany. Graddol (1996) identifies the following officially recognized languages within member states: Alsatian, Asturian, Basque, Catalan, Corsican, Frisian, Galician, Ladin, Luxembourgish, Occitan, Sardinian, Scots Gaelic, Welsh. Furthermore, indigenous EU communities include speakers of Albanian, Aragonese, Breton, Cornish, Franco-Provençal, Friulian, Karelian, Lallans, Macedonian (Greece), Manx, Polish, Romany, Sarnish, Croat, Slovene, Serbian, Turkish, and Vlach.

For many of these groups, language has been the central plank of identity, and as Wright (2000: 187) shows, the European Community – especially its Parliament – has from the late 1970s been active in raising ethnic minority language rights, with a committee set up in 1983 to examine how education, media, and cultural initiatives could reflect regional language more effectively. Crucially, however, linguistic revitalisation needs more than mere acquiescence to the demands of language campaigners:

> For most families there would have to be clear economic/social/political benefits accompanying a return to a language which had often been abandoned because it hampered social promotion and the realisation of career ambitions, and blocked interaction with the state and wider civil society.
>
> (Wright 2000: 190)

What we see emerging from the so-called 'neo-medieval' perspective on European integration is the fact that bilingualism is now emerging as an important factor of European society. This may mean, therefore, that it may be as feasible for Basques to learn English than Spanish. They may decide to communicate to each other in their 'local' language, but engage with the 'global' in English.

Functional economic spaces and structural funds

In many parts of Europe, territorial organisation along nation-state lines has not kept pace with economic trends. It is clear that, for example, Catalonia and Languedoc-Roussillon, neighbouring each other on the Spanish and French border, might have as many shared concerns as with their respective capitals hundreds of kilometres to north and south. Areas that share strong infrastructural links – such as the newly created Malmo–Øresund bridge link or Kent–Nord–Pas de Calais Transmanche (linked by the Channel Tunnel) – have a rationale in overcoming cultural or linguistic differences in pursuit of joint prosperity. For example, Keating (1998: 9) suggests that 'an economic definition of a region would focus on common production patterns, interdependencies and market linkages and labour markets. A broader functional definition would add patterns of social interaction, including leisure, recreation and travel patterns'. To expand this, regions might be 'frozen' for statistical purposes, their territory used as a container for measuring economic activity. But much of this activity might be invisible, unless a more dynamic map of cross-border flows, commuting, house-buying, etc. was also created.

It is clear, though, that while national governments can produce macroeconomic policies that fit with general trends in the global economy, the regions can 'fine-tune' this in a way that makes a world of difference to whether foreign corporations invest, or whether domestic producers have the capital, labour force, and skills to export successfully. This, microecononomic or supply-side policy, is one of the great rationales of regionalism. Regional governments or development agencies have greater sensitivity to local market conditions (in terms of, say, commissioning research on labour markets, or actual 'on the ground' knowledge of environmental issues or the labour needs of big firms) than unwieldy central bureaucracies. As such, they are ideally placed to stimulate supply-side measures such as site provision for inward investors or regional training programmes, while the 'macro' issues of employment or monetary policy are increasingly addressed in Brussels.

For a Single European Market to be successful required the creation of a relatively even economic space where, for example, peripheral regions (known as 'lagging' regions in EU-speak) were given subsidy from the likes of the European Social Fund, Cohesion Fund, or European Regional Development Fund. This 'carrot' has undoubtedly worked to foster European consciousness among regional policy-makers and has given a sound economic rationale to arguments for greater regional autonomy. However, the increasing funds being channelled to post-communist states awaiting membership is likely to dilute this factor. Nonetheless, many regional and local authorities have set up offices in Brussels both for lobbying and inward investment purposes. Above all, the Commission has called for a new, post-national territorial logic where macro-economic trends and functional trading patterns are given priority funding.

This logic was expressed in a number of European Union strategy documents, notably *Europe 2000* and *Europe 2000+,* and more recently the European Spatial Development Perspective (ESDP) (Commission for the European Communities 1991, 1994, 1999, respectively). In *Europe 2000,* the community's territory was divided into eight macro-regions, with boundaries drawn in complete contrast to existing national shapes. Thus the 'Alpine Arc' identified the area stretching from the north of Italy, southwest of France, western Austria, southeastern Germany and Switzerland as possessing strong trans-national linkages and shared policy issues. The 'Continental Diagonal' linked the primarily agricultural inland areas of France and Spain. And the 'Atlantic Arc', running from the Scottish Highlands down through Wales, Cornwall, Ireland, and the French, Spanish, and Portuguese Atlantic coasts, reflecting the apparent similarities in terms of maritime economies, coastlines, and cultures.

As Wise (2000) has decribed, this latter area spawned a permanent commission which sought, through lobbying and joint working, to reverse the perception that these areas were 'peripheral' to the EU. Threats such as the improved accessibility of the territorial 'core' areas, the competition posed by the imminent accession of countries of East and

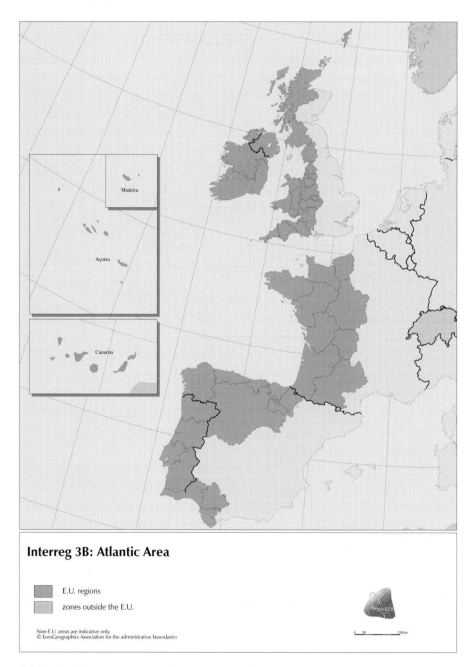

Interreg 3B: Atlantic Area

E.U. regions

zones outside the E.U.

Non-E.U. areas are indicative only.
© EuroGeographics Association for the administrative boundaries

0   50   250km

Figure 9 *Map of Atlantic Arc. © European Communities 1995–2002.*

Central Europe, and the potential centralisation caused by the creation of a single European currency helped give some of these geographically disparate regions common cause. Yet Wise argues that despite a limited number of concrete activities, the Arc was more of a mirage than a reality, for six reasons: first, most regions had stronger ties to their inland, national neighbours; second, attempts to forge a common 'Celtic' cultural identity found few takers among the 'the Unionist population in Northern Ireland, the Basques of northern Spain, the wine-growers of Bordeaux, and the dairy farmers of Devon' (Wise 2000: 879); third, the sight of 'Spanish' fishermen in 'British' waters prompted more rivalry than co-operation; fourth, the shared socio-economic character-istics of rural coastal areas such as Galicia had little in common with the likes of industrial Liverpool; fifth, the most active participants in the Arc tended to be near neighbours, as with regions in Spain and France; sixth, Wise concludes that the overwhelming reason for the structure's existence was to profit from EU Structural Funds.

In short, the creation of a transnational, inter-regional form of governance from on-high remains difficult to achieve. There are many areas where, on a smaller scale, such transnational projects are working successfully. But even in these cases, the resurgence of 'particularist' and strongly ethnicised cultural identities remain more of a barrier than an aid to a joint European consciousness.

High-tech regions

This apparent argument for greater regional autonomy was given further impetus by the apparent success of certain key regions in the European economy of the 1990s (Dunford and Kafkalas 1992). These areas – Emilia Romagna or Baden-Württemberg, for example – were held to possess localised knowledge and relations of trust that were not present in the more abstracted global economy. Indeed, it appeared that it was in these regions that trans-national corporations preferred to invest, particularly in the higher-skilled levels of their operations, due to advanced training programmes. Why did this happen? Both in the German case, in the likes of Bavaria and Baden-Württemberg, very interventionist regional governments encouraged the private sector to contribute to science parks and specialist research institutes, easing the transition to a post-Fordist economy (see Amin 1994 for an overview). In France, the central government planning agency (DATAR) initially realised the need to decentralise and spread economic activity throughout the country due to fears that Paris was sucking the rest of the national territory dry, and sponsored high-tech poles in the south of the country in the likes of Toulouse (Airbus and Ariane space rockets) or Lyon (the agricultural and financial powerhouse of Rhônes-Alpes) (Castells and Hall 1994). It should be borne in mind that the siting of nuclear power stations in Brittany, part of this same regional policy, fomented regionalist backlash here (Ardagh 2000). Furthermore, such high technology futures are not open to all regions. In many cases, the drivers of a Europe of the Regions may be pursuing a 'bourgeois regionalism' which can be seen 'as clear special pleading on behalf of regions that had already concentrated in themselves a disproportionately large amount of the investment and the organisational structures of European capitalism' (Harvie 1994: 5). Massey *et al.* (1992) provide a well-needed critical analysis of the presumptions of a high-technology strategy in the context of the southeast of England.

Global or local? Regionalism as a political project

What was interesting about these developments was not that they were seen as strictly new, but rather that they reasserted identities that had been subdued in the heyday of national capitalism (the post-war decades of the 1950s and 1960s, particularly). Post-war Europe could, if written from a regional viewpoint, present a very distinct perspective on European identity (as Harvie 1994 outlines in *The Rise of Regional Europe*). In many ways, they came to represent what some commentators called a 'neo-medieval' Europe, where great powers (cf. The European Union) co-existed with religious and locally organised states.

> It is a dream of many Europeans that the continent may be reverting to a medieval pattern. A common language (probably American-English) in place of Latin; a common scientific culture in place of the common theology; the mass proletariat and trade unions splitting up into more specialized skills, like medieval guilds . . . a Europe of the Regions would be the climax of that trend: the map of the continent would begin to look like a map which hangs in my hall, of Europe 'between Ancient and Modern Geogaphy', with WEST SEXIA, AQUITANIA and LOMBARDIA in big letters, and BRITANNIA, FRANCIA and ITALIA lying across them.
>
> (Sampson 1968: 432)

The dynamic of globalisation is thus creating a kind of post-national scenario, where local identities are expressed while simultaneously interacting with supranational actors. There are numerous examples, some of which I explore here: the opening of the Bilbao Guggenheim art gallery; the fusion of Northern Italian economic prosperity with appeals to pre-modern ethnic identity and history; the Braveheart phenomenon in Scotland; and the 1992 Barcelona Olympics.

Box 3   Particularist pragmatism: the Bilbao Guggenheim

The first example I want to draw on illustrates the intent by a regionalist government – the Comunidad Autónoma Vasco – to come to terms with the practicalities of retaining political-economic autonomy and vibrancy in a global economy, with the possible loss of cultural identity. Challenged by the separatist terrorism of ETA on the one hand, and with the threat of having its powers restricted by the Spanish central government on the other, the moderate nationalist Partido Nacional Vasco threw itself into modernising its key city, Bilbao. Its solution was to enter into negotiations with one of the most famous of the world's art institutions, the Guggenheim Foundation. With its

(a)

(b)

Figure 10a,b *Bilbao Guggenheim. Donald McNeill.*

flagship gallery on Manhattan's Fifth Avenue too small to display its huge collection, it sought to expand around the world through franchising (a similar management structure to that of McDonald's). As such, this potential 'McGuggenisation' of the Basque country was controversial (I discuss the argument presented below more fully in McNeill (2000)).

Opened in October 1997, the Bilbao Guggenheim art gallery was notable for three reasons. First, as a visual masterpiece and architectural oddity, Frank Gehry's glimmering, titanium-dressed spectacle of curving walls, window shards, and abstract motifs caught the attention of the world media, popular as well as architectural. Its visual distinctiveness made it an instantly recognisable piece of place marketing, featuring in a James Bond film as well as forming a backdrop to numerous television adverts. Second, its location in Bilbao – a hitherto rather unfashionable industrial port – reflected a determination on the part of the Partido Nacional Vasco (PNV), to place itself on the global map (understood, perhaps, as the grapevine of images and gossip conducted through the international business community and its associated support services of airline magazines, hotels, and convention centres). Third, it represented the most ambitious and entrepreneurial piece of business acumen on the part of an art foundation hitherto seen. Under Thomas Krens, the Bilbao Guggenheim was only the first of a process of franchising moves that has brought Guggenheim museums to Berlin and Las Vegas.

The negotiation behind the Bilbao Guggenheim is a perfect example of the apparent by-passing of the nation-state by regional or urban political elites. While the Basque case is rather unique in terms of the extent of their financial autonomy from the Spanish state, their willingness to proactively negotiate a deal with a global cultural institution (with an *explicit* avoidance of Spanish mediators) demonstrates both the confidence of such 'local' actors under conditions of globalisation, and an awareness of the strategic nature of such negotiation. It was clear that the Basque government was a weaker partner in the negotiations with the Guggenheim, for a variety of reasons (the fear of losing the franchise to competitor cities, a poor knowledge of the global art world, the sizeable sum of money paid for use of the foundation name and curatorial services, and for the building (Zulaika 1997)). And yet, as the PNV-appointed director of the museum, Juan Vidarte, put it: 'With this unique space and this important collection, we can be playing a role in the periphery that we could not do otherwise ... To play in this league, you have to be associated with someone in it. Otherwise, it's hard to get there' (cited in Cembalest 1997: 64). It is this sensitivity to – yet refusal to feel excluded by – geographical peripherality that marks out the Basque case.

So while the arrival of the Guggenheim could be seen as a process of cultural colonisation or imperialism (funds were diverted from indigenous cultural sectors to help pay to bring the American collection to Bilbao) it can also be seen as an attempt to escape the Spanish political domination that had been such a part of Basque life during the Francoist dictatorship. Simultaneously, it can be viewed as an attempt by the moderate nationalist PNV to distance itself from the 'primordial' violence of ETA separatists, while advertising Basque difference to the world.

Inventing the region: the Lega Nord

The 1990s saw the re-emergence of numerous far-right political parties. While Fraser (2000) explores the nature of the likes of the racist National Front in France, he also identifies a grouping of what he calls 'the new lost tribes of Europe', a rather diverse, but no less interesting, set of regionalist movements. They are almost all 'populist', in the sense that they operated by 'some sort of direct appeal, visceral in nature, and bypassing reason. Populists weren't necessarily racists, socialists, or anti-Europeans, though they might be all or any one of these things' (Fraser 2000: 254). Almost all combine a strong ethnic sense of community (e.g. shared language, history, or landscape and sporting symbolism) with economic prosperity. This latter is important: almost all drew strength from their attacks on the capital cities of their nation-states, or Brussels as EU capital, or both. Here, parties such as the Vlaams Blok (Flanders) and the Austrian Freedom Party, led by Jörg Haider, which entered coalition government in Austria, have enjoyed moderate political success.

The most striking example of this 'region-building' comes from Italy, from the Lega Nord, and its invented homeland of 'Padania', the region straddling the River Po. The Lega emerged as an important movement in the general elections of 1994, alongside Silvio Berlusconi's Forza Italia. Indeed, the ability of FI to govern was only thanks to the coalition of two other parties. One was the neo-fascist National Alliance, the other the Lega Nord (LN). The LN's disastrous showing in the 2001 general election, where many of their clothes were stolen by Berlusconi, suggests that they may be a transitory phenomenon. Yet Berlusconi's incorporation of their leader, Umberto Bossi, into government, means that LN are still an influential force. Regardless of their electoral fortunes, however, a study of the LN reveals that many of the trends discussed in the last chapter – political branding, the power of national symbolism, the creation of 'others' – was exercised with considerable skill by Bossi and his supporters. I examine this in three ways.

First, the LN exemplifies the *'othering'* of regions within the nation-state. Italy – as in Spain, Belgium, and the UK and, to a lesser extent, France – has always been characterised by a crude north–south divide (Giordano 2000). It is often said that Milan is closer to London than it is to Palermo, and it is also popularly noted that Sicily is as African as Milan is Swiss, whatever such stereotypes may mean. Agnew reflects that:

> For some years I have been fascinated, if somewhat perplexed, by the ease with which Italians and others account for some particular feature of 'Italian life' in terms of Italy's generic 'backwardness' or 'immaturity' in comparison to other European countries. More often than not, invoking backwardness appears to involve situating the phenomenon in question (Mafia, government deficit, economic disparities between north and south) in a simple temporal backward-modern couplet that all who read or hear intuitively understand as meaningful.
>
> (Agnew 1997: 23)

This polarity has a clear urban dimension, expressed particularly in the discursive and political battles over the right to be the 'true' capital of Italy:

> On the one hand, Milan was the true capital because Rome was not – the myth was a negative assessment of the contribution of Rome and the south to Italy's industrial progress. Milan was modern, industrious, hard-working, honest, productive; Rome was corrupt, unproductive, lazy and pre-modern. Rome was the political capital, Milan the real driving force of the nation, its moral heart. Yet the myth was also a celebration of these values in themselves – a series of character-traits and concrete realities pertaining to the Milanese worker, entrepreneur and industrialist – once again, modernity, hard work, thrift, legality, the self-made man.
>
> (Foot 1999: 405)

However, while the self-made men such as the tyre-manufacturer Pirelli came to give substance to Milan's 'moral' claim to leadership in a divided Italy, and Milan's 'entrepreneurial' work, work, work image mushroomed in the industrial boom of the 1980s to a city of 'Europeanness', new wealth, footballing success, and dominance of the high fashion world, the Milanese image was shattered by the discovery of a vast network of corruption involving substantial numbers of these self-same self-made men. Milan became known as 'Tangentopoli', or 'bribesville', many of its leading political and business figures, including ex-mayors, arrested and charged. While the scandal was quickly discovered to stretch to many other major cities, including Rome and Naples, it was Milan that was to remain with a highly dubious image even after 1993. Yet in the hands of unscrupulous, yet highly skilled, politicians such as Bossi, North versus South makes good, clear electoral politics. As we saw in Chapter 1, Central Europeans may be distancing themselves from the Balkans and 'Eastern' Europe through the use of dualisms. In Italy, the same process is at work.

Tambini (2001: 123) identifies a series of dualisms where the backwardness thesis is carried out, focusing on the differing relationships between North and South and the Italian state (see also Dickie 1996; Lupo 1996). In particular, the North is held to be 'innately' entrepreneurial, industrialist, and hard-working, by contrast with the lazy, predominantly agricultural, southerners. The North is, crucially, seen as net tax-payers to a Southern state apparatus that funnels off resources to a Mafia-controlled South. These dualisms arise from a perceived sense of injustice within the Italian political system dating back decades, where central government in Rome is discursively constructed as a centre of corruption, a breeding ground for Mafia activities in the South, a source of welfare subsidy for lazy southern farmers, paid for by Northern taxes. Such an argument – while crude – plays to the prejudices of a certain section of Northern voters. As such, Bossi and his allies were able to mobilise a set of myths that the Northern League could turn into a political package. Yet to do this, the LN needed a stronger brand, or set of traditions, that during the course of the 1980s and 1990s began to emerge in party organisation and electioneering.

So, second, the LN became a textbook example of the *symbolic invention of place identity*. From discussions in the previous two chapters, it should be clear that place images are socially constructed. Stereotypes and traditions have no independent existence outside the speeches of politicians, the popular knowledge of voters, the manipulation of symbols, and so on. European regionalism is no exception to this. The LN were able to construct a number of ethnically and racially based discourses with which to animate their potential voters. We could summarise these as follows.

- **Lombard tradition**

  > The League's was a post-modern history: there were no authorities to refer to, neither references nor footnotes, only haphazard constructions of whatever story fitted the desires and the political opportunities of the day.
  >
  > (Tambini 2001: 117)

  This idea of a selective construction of historical identity is not confined to the League, but their confection of an identity based on dubious readings of historical events was central to their project. The clearest reference was to the medieval Lombard League, which united 20 Northern cities and fought against outside invasions in 1176. As Tambini puts it, 'This historical episode, perfectly genuine by all accounts, was selected and inflated by League propagandists into a claim of a Lombard tradition of resistance to centralising foreign imperialism' (2001: 117). Similarly, Venetian traditions of maritime dominance were fused with the Lombard identity when the idea of a larger pan-Northern movement was conceived.

- **Northerners as hard-working Calvinists (ie anti-Catholic)**

  Drawing sharp contrast to the southern part of Italy (including Rome), Bossi developed a theme similar to Max Weber's protestant work ethic as a means of tying the idea of efficiency and hard work to Lombard ethnic identity. Part of this was rooted in anti-Catholic identity (although Catholic sections were incorporated in the League).

- **Northerners as Celts**

  > In Italy, more than in other Western European nation-states, there is a diffuse conception that there are phenotypic differences between the populations of the different regions, and particularly between North and South. Lively, if confused, conceptions of the various so-called 'ethnies' (the Celts, Etruscans, Venetians, Romans and so on) that inhabited the peninsula in pre-modern times lurk behind many stereotypes.
  >
  > (Tambini 2001: 110–1)

  Such a crude idea of biological difference was gleefully seized upon by League ideologues. In the search for a unifying focus for Northern identity, the Celtic myth was a dynamic and romantic resource for the 'imagining' of the North. As Tambini (2001: 111) continues, 'the Celtic argument fitted perfectly with the Northern League's slogan, '*Più lontano da Roma, più vicino all "Europa"*' (Further from Rome, closer to Europe). The Roman-Celt opposition is familiar to any reader of Asterix'.

- **Anti-immigrant xenophobia**

  The economic dimensions of the League – a powerful call for an end to subsidising the South through taxes – included a strong anti-immigrant focus (where Southern

Italians were equated with North Africans). As well as promoting a sense of Lombard racial 'purity' or identity, Bossi was rumoured to have forged links with Austrian neo-fascist Jörg Haider, who had been included in Austrian government since 1999. Acting together, these two could form an Alpine neo-fascist block. Here, the idea of Rome (representing central government) as being a black hole where Northern tax revenues were shovelled, disappearing into corrupt, mafia-controlled public works projects was combined with the counter-argument, that immigrants from the South were 'invading' Padania, and stealing Northern jobs. In the process, the League were happy to liken the South of Italy with Africa, thus 'othering' them to the 'European' identity of the North.

Third, under Umberto Bossi, we can see the importance of *performance politics and demagoguery* in this brand of politics. Bossi is one of a number of populist leaders – such as Haider in Austria, Le Pen and Mégret in France, the late Pim Fortuyn in the Netherlands – who have been able to mobilise a substantial number of followers and voters with a blunt, simple language that contrasts sharply with the rhetoric of professional politicians. When combined with an appeal to place identity, this has a potent political impact. Such rhetoric is not confined to typical media outlets, however. As Tambini (2001) suggests, the distaste that many media commentators showed to the League's xenophobic and crude programme meant that the 'oxygen of publicity' was often denied them. In response, the League began to develop their own publicity-grabbing festivals. At the Festa on the Po in 1996, Bossi drew water from the Po river as signifying the invented region's ecological roots. A regatta in the Venice lagoon in 1998 further projected the secessionist call to an international audience, as Fraser describes:

> Gusts pushed clouds in our direction across the Rialto, blowing the banners of the Padanians to and fro like Chinese dragons. Sirens and bells filled the air with bland, friendly noises. You could buy Padanian banknotes, passports and even blood-donor certificates in a series of stalls grouped along the rainswept Riva degli Schiavoni. They were inscribed with the likeness of Umberto Bossi ... A large pontoon held a dais for dignitaries and well-wishers, and I stood behind a noisy group of Padanians who were swaying tipsily ... As I became habituated to the football chants, however, I caught a refrain. 'Don't jump if you're an Italian,' the crowd shouted, and they were jumping. And now I was alerted to a change of mood by the swelling chords of Verdi's chorus of slaves and the waving of many small pieces of paper. This was the celebrated oath of loyalty, sworn on the sacred honour of the crowd, and it pledged the huddling masses not to tolerate the existence of Italy but to fight for a free Padania. I looked around for signs of invasive irony and realized that these were real tears. Many were sniffling into their handkerchiefs or conspicuously striving to overcome the growing lumps in their throats.
>
> (Fraser 2000: 260–1)

The importance of the Venetian landscape was central to the attempts to forge a larger entity out of the smaller regional leagues, and create a whole 'Northern' region including Piedmont and Lombardy. Indeed, it was little surprise when in May 1997, a splinter group of radical nationalists stormed the bell tower of St Mark's Cathedral, proclaiming the independence of the Republic of Venice from Italy. This group – the so-called 'Venice Eight' – were armed, hijacking a ferry, and driving across Piazza San Marco in an armoured car (Plant 2003). While the League distanced themselves from the action, many Italians blamed them for awakening the spectre of separatist terrorism seen in Ireland, Corsica and the Basque Country.

So we can see that Padania, or Flanders, or Carinthia, may represent rather worrying throwbacks to fascism and the ghosts or pre-war Europe. Yet to leave the story there would be to do the regionalist resurgence a disservice, for there are many progressive movements operating at this level too. Growing up in a region or nation where you have no significant political representation, listening to people in power from parts of the nation with accents very different from your own, suffering from economic policies that benefit financial sectors when you live in an area of manufacturing, these are all little grievances that easily breed resistance to an over-powerful centre, leaving even the most rational citizen happy to raise two fingers to the remote politicians who present themselves as models of sensitivity. And so, a majority of the Basque electorate – suffering under a wave of nihilistic ETA violence – rejected Spanish prime minister Aznar's Partido Popular in the 2001 regional elections. So did the Scottish electorate again and again under the Conservatives in the 1980s and early 1990s, when many parts of England were returning right-wing politicians. What is important – and this is an unfortunately muffled argument – is that the regional dimension has no clear 'good' or 'bad' ethic to it. It can be exploited and used politically, both to enhance democratic political representation and to simultaneously encourage a cultural particularism that may exclude outsiders.

The Braveheart phenomenon

> Satellites, cable and commercial television, computers and VCRs, all heightened the danger of Europe's 'cultural suicide' (in the melodramatic phrase of a bureaucrat in Brussels). But the spread of the new technology did not, by itself, signal the death of a 'European' consciousness. The end would come when Italians and Germans, Greeks and Norwegians, the Poles and the Portuguese, each decided that their private fantasies and favorite folktales were not as compelling as the intricate plots of a John Grisham novel or the larger-than-life heroes in The Terminator and on Baywatch. No longer willing or able to invent their own stories, people in Europe would rely in the future, far more than they had in the past, on the American media for the emotional sustenance that their local cultures once provided but now could not.
>
> (Pells 1997: 269)

Such apocalyptic visions of cultural imperialism are far from exaggerated, but they ignore a very important aspect of mediated transmission – that all cultural products are received and understood within local frames of understanding.

An excellent example of this is provided by Tim Edensor (1997, 2002, see also Morgan 1999) in his discussions of the film Braveheart, a Hollywood 'blockbuster' released in 1995. The film – directed and starring Mel Gibson – tells the story of William Wallace, a Scottish medieval warrior chief who won notable battles against occupying English forces. The content of the film was, predictably, criticised for numerous inaccuracies in its narration of the Wallace story, as well as for perpetuating a long-running romanticisation of Scotland as a land of mists, castles, and hard men. Yet more interesting than the critique of the firm's plot was its public reception. As the fifth most popular UK box office hit of the year, the film's apparent anti-English sentiment and archaic wallow in Scottish romanticism had two very important social effects, one political, one economic-cultural.

As a political event, the film's release coincided with the growing demands for independence made by the Scottish National Party. As Edensor notes, the party campaigned outside cinemas showing the film, and distributed leaflets 'in the form of reply-paid postcards':

> On one side was an image of Mel Gibson as Wallace and 'BRAVEHEART' in large capitals, along with a text, culminating in the words: 'TODAY IT'S NOT JUST BRAVEHEARTS WHO CHOOSE INDEPENDENCE – IT'S ALSO WISEHEADS – AND THEY USE THE BALLOT BOX'. On the other side is the slogan 'YOU'VE SEEN THE MOVIE … NOW FACE THE REALITY'.
>
> (Edensor 2002: 151)

With a subsequent surge in opinion polls and membership applications, it could be argued that the Hollywood product mobilised a culturally specific response, which some commentators called the 'Braveheart effect' (cf. The 'Guggenheim Effect').

The second major impact was in reaffirming the beauty of the Scottish landscape, a key sector in the regional economy. The areas around Stirling – the location of the film and the site of the key (real) battle – brought in millions of pounds in tourism revenue, leading some to worry of Scotland as theme park, a Highland fantasy propagated by Hollywood.

Many viewed the film with disquiet. There were several arguments raised about the film – some worried about modern nationalist movements praising a medieval warrior, others about the 'othering' of the English, and still others criticising the deliberate bending of historical facts (in a manner similar to the Euro Disney debate discussed in Chapter 2). As Edensor (1997: 188) puts it, 'the image factory of Hollywood has recast the myth, partly in accordance with its requirements of action sequences, characterisation, and narrative, to appeal to an international audience. This appears to confirm the … processes that disembed social memory from place'. So here we have a case where one of the key global institutions – the Hollywood film industry – takes a definite historical event and reworks it, which is then broadcast and *appropriated* by 'local' actors to seek political capital against the nation-state. The parallels with the Bilbao Guggenheim should be clear – to a certain extent, Braveheart was indigenised by Scots, a local story brought to life by a distant-by-origin, and slightly fantastical, cultural flow.

The 1992 Olympics: Barcelona versus Catalonia?

The Olympics are, along with the World Cup, perhaps the most important global media spectacular that there is. Watched by millions around the globe, such a reach has meant that transnational corporations have seen the Games as the perfect medium (and metaphor) to celebrate their mass economies-of-scale branding and products. From Nike to Kodak to CNN, they are all there, heavily involved in broadcasting, filming, or dressing the athletes, along with countless others who buy the advertising space offered at the stadia, sponsorship deals and commercial breaks.

Gaining the Olympics is thus a major coup for any government. 1980 and 1984 (Moscow and Los Angeles) were superpower extravaganzas, Seoul in 1988 was a shop window for the 'Asian Tiger' of the South Korean economy. Yet Barcelona 1992 was different. The bad press surrounding the previous events – boycotts and doping scandals – had discredited the Games, and it is was undoubtedly the softer profile offered by Barcelona that encouraged many to support it. Yet this didn't mean that the Games were without political tension, and the story of their staging reveals many of the conflicts that run through regionalist identities. For, as I and others have discussed elsewhere (McNeill 1999b; Hargreaves 2000), the 1992 Games could be presented as a battle *between competing visions* of Catalan identity. Here was a case of cities *versus* regions, operating strategically both against and with the nation-state.

There were two main actors in the conflict. On the one hand, there was the president of the regional government (Generalitat), Jordi Pujol, whose party Convergéncia i Unió (CiU) had long pursued greater autonomy for Catalonia and, particularly, its linguistic identity, engaging in conflict with central government, if necessary. On the other side was the mayor of Barcelona, Pasqual Maragall, whose social democrat party (Partit dels Socialistes de Catalunya) represented much of the city's liberal middle classes and, importantly, working class immigrants from other parts of Spain. For them, the linguistic issue was important but not fundamental. Maragall represented this as a 'Catalunya aberta' (open, outward-looking Catalonia) as opposed to a 'Catalunya tancada' (closed, intro-spective Catalonia). So, while Pujol put Catalan language (in education and broadcasting) at the forefront of his programme, Maragall sought to represent Catalonia through its major city. Both sides had a very different conception of Catalonia. For the PSC, urban modernity (expressed in the use of modern architects and the staging of avant-garde cultural events) centred around Barcelona was seen as central to a cosmopolitan Catalanism. By contrast, the CiU under Pujol sought to preserve the 'timeless' essences of Catalan culture where possible, embodied in the work of Gaudí and Catalan rural identities.

While the history of confrontation between the two sides is complex – they are, after all, pursuing the same goal of enhancing Catalan identity and autonomy, but in different ways – both fully exploited the staging of the Games, as follows:

- The use of a global media event to project their respective territories (Barcelona and Catalonia) onto a world stage. The Generalitat spent substantial amounts of public money in taking out adverts in the world's leading newspapers and news magazines, such as the *Financial Times*, *New York Times*, *International Herald Tribune*, *Der Spiegel*, *La Repubblica*, and many more. The city council were able to bask in the coverage given to the city's name in constant references to the sporting events, as well as associated media interest in the city itself, thus boosting tourism prior to and after the event.

- The use of the protocol at the opening and closing ceremonies to establish the legitimacy of Catalan autonomy after the repression of the dictatorship. The presence of the Catalan flag and national anthem alongside those of Spain were clear evidence of this, with the attendance of the Spanish royal family legitimising the post-Franco settlement.

- As the Guggenheim would later do for Bilbao, Barcelona city council used the Olympic one-off investment for the Games from private investors and central government to modernise the city's infrastructures. This also included the building of an entire new residential district, the Olympic Village, which was designed to retain high spending residents in the city's ambit.

However, these achievements took place against a backdrop of political bickering, disagreements over funding, electoral manoeuvring, and grassroots campaigning.

So, the 1992 Olympics illustrated two things. First, the willingness of regional elites to use global spectacle to advertise themselves and the cultural attributes of their territorial units (be it city or region) in a way that distinguishes them from the nation-state. Second, it shows that political conflict can exist *within* regionalist movements, even if superficially they share the same goals.

Conclusions

So, a number of factors and trends have contributed to a growing sense of a 'Europe of the Regions':

- The idea of European integration as an 'umbrella' for groups seeking greater cultural autonomy but fearful of the effects of economic autarky (ie. self-sufficient isolation) or separatism from national markets.

- The popularity of regional economic strategies, where science parks and infra-structure are combined with a high standard of living to encourage transnational corporations to invest in their region.

- The formal political representation and economic support offered by EU member-ship and structural funding, particularly through the Committee of the Regions and the resources of the European Regional Development Fund.

- The rise of increasingly confident regionalist parties, some of which co-operate in transnational lobbying activities, and which are increasingly professional in their organ-isation and voter appeal.

These regionalist parties (and associated social movements) are building upon existent cultural identities that may have been submerged by centralised nation-building projects. In particular, the revival of minority languages in education and broadcasting, the celebration of alternative readings of history, lingering senses of political oppression (e.g. Basques and Catalans under the Francoist dictatorship) or a sense of grievance at central government economic policy and political legitimacy (e.g. Scotland and Northern Italy) have been exploited and packaged (e.g. fusing many of these factors) by parties such as the Scottish National Party or the Northern League. Furthermore, while there is evidence of a tendency to nostalgia and 'looking backward' to pre-modern ethnic and national identities, many regional governments are now operating at a 'global' level, to by-pass the nation-state in its strategies. Thus the likes of Bavaria or Euskadi may deal directly with transnational corporations, and may combine religious or linguistic traditionalism with high-tech science parks and avant-garde architecture. The 1992 Olympics case study suggests that we are not witnessing a straightforward passing down of powers from nation-state to regional or city government, and that there are many examples of conflict between regions and cities.

So, it may be misleading to speak of a *regionalisation* of Europe as opposed to a more general *reterritorialisation* of the continent. In other words, a shift from the one-dimensional map of Europe as having fixed borders to one in which city-based, regional, national, and European scales of action are fluid. Inter-scalar politics — crudely called 'glocal' — may be the most accurate way of viewing this. This fits with a 'neo-medieval' conception of Europe (Bull 1977), one in which regions may negotiate with corporations, but only in the framework provided by the nation-state, where national languages such as Spanish or French might be superseded by a bilingualism of Breton and English or Basque and English, and where television and cyberspace are more important means of communication and community formation than formal educational institutions. And as I now discuss, *cities* are emerging as important players in a reterritorialised Europe.

4

A Europe of the cities?

> One of the city's most evident geographic advantages is that it offers two types of proximity. One is territorial proximity, in terms of density and social neighbourhood. But the city also offers proximity in networks in relation to other cities. Thanks to advanced means of transportation and communication, people, buildings and institutions are within reach and easy access despite physical distances. The nodes in institutional networks attract each other. The management of industry, finance, research, interest groups and public administration is concentrated in cities, as are clusters of specialized services. The media broadcast news, culture and entertainment from large cities. Cities represent the dense environments that throughout history have provided meeting-places, crucial to renewal and artistic creativity.
>
> (Jönsson et al. 2000: 157)

We have seen that nations and regions usually have a clear set of identities around which political projects can be based. But what about *cities* in Europe? There are two issues to confront at the very outset. First, it is not even clear that a 'city' exists for some social scientists, who prefer to speak of *urban* processes or systems, and with good reason. Many theorists argue that cities should be discussed *relationally* – that is, in terms of their relations with other cities and places, rather than being seen as sealed containers of social life. For Ulf Hannerz, 'cities . . . are good to think with, as we try to grasp the networks of relationships which organise the global ecumene of today. They are places with especially intricate goings-on, and at the same time reach out widely into the world, and toward one another' (1996: 13). Second, contemporary anglophone urban theory is often lacking in concern for the specificity of place, with 'universal' theories or models being developed from paradigmatic cities – Chicago, or Los Angeles, particularly. Part of the thrust of globalisation theory is that places are losing their differences and are becoming increasingly homogenised.

Here, I want to counter these approaches. First, for all the efforts of social scientists to dispel the mythology of the city, within wider public discourses – such as the mass media, or in everyday speech – people continue to refer to cities as having a uniqueness, a life of their own, a set of characteristics that mark them out as being different. Within the arts, the city is often the centre of ideas of human expression. As Carl Schorske (1963) has described, for key thinkers from Voltaire to Adam Smith to Karl Marx and Charles Baudelaire, the city has been central to the evolution of modern cultural identity, a civilising force, fundamental to the project of European rational thought. Simultaneously, it has been the source of angst, where the 'city of vice', of a degraded urban life associated with nineteenth-century industrialisation, has provoked utopian thoughts both archaic and futuristic (Schorske 1963: 104). And at a more popular level, there is a mythical and essentialised attachment to the stereotyped inhabitants of the city, which some commentators have exposed with great skill (see Spring 1990 on the Glaswegian hard man, Robb 1998 on the theatrical, fatalistic Neapolitan, or Foot 1999 on the hard-working Milanesi). The myth of ancient Rome and Athens pervades European identity. The architecture of Florence, Edinburgh, and Barcelona gives focus and coherence to these place-ideas. City identities are by their nature mythicised, and play into social practices and actions.

In this chapter I argue that we can see Europe through its cities, but that we can only do so by holding *together* these two, rather polarised, ideas of cities as being defined by flow and fixity. This tension – between the relational city and the essential city – runs to the heart of the chapter. It is driven primarily by new work on globalisation and trans-nationalism (e.g. Smith 2001) and the impact of globalisation infrastructure (Graham and Marvin 2001), but also by the substantial evidence that, rightly or wrongly, cities are often viewed as being unique and having lives and biographies of their own. And these apparently opposing trends may in fact be linked. Over the last 20 years there is little doubt that political elites have *refound* their civic identities, have sought to embellish or rejuvenate their cities. This urban renaissance – or rebirth – is (as the term implies) built upon particular histories, and mythical essences. Yet this is done as a means of mobilising their cities within a Europe defined by economic transactions, the flows of tourists and migrant workers and commuters between cities, by the development of air routes and rail networks and motorways, by conference circuits, by the economic geography of movement of goods, by the advertising industry and the flow of images of cities from place to place.

So, this chapter addresses some of these tensions. First, it briefly discusses the theoretical understanding of the city as being essentialised – possessing some kind of unique and describable 'character' – and *at the same time* relational, in other words, *defined* by its links and interactions and competition with other places. Second, it considers the existence of an 'essential' city, both in terms of an idealised city-dweller, such as the Parisian, Neapolitan, and Glaswegian, and the in the sense of cities having biographies, of being like people. Here, I briefly flag up the literature that suggests these denizens have been forced out or dispossessed by a new collection of city dwellers, and in turn that cities have lost their distinctive identities. Third, it examines the growing work on transnationalism and cities, looking at the notion of cross-cultural flows of people and ideas that destabilise the notion of, say, Stockholm as a Swedish city, Lisbon as a Portuguese city. Fourth, I discuss the role of mayors – who both essentialise and relativise their cities simultaneously – given the emerging profile of this latter group in the New Europe. Finally, I consider how these identities may be being 'disembedded', where the impact of television on football might be seeing the selling of place-based essences (e.g. Barcelona, Manchester United) to global audiences, while distancing them from their local origins. And as a further example of this, I profile the position of the

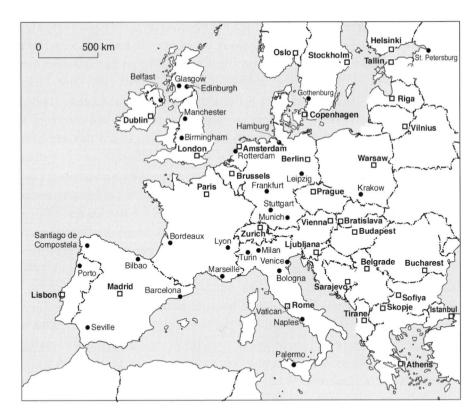

Figure 11 *Map of a Europe of the Cities.*

Vatican as being at the centre of networks of pilgrimage, flows of ideas, and belief systems, and at the same time possessing a mythical essence based on St Peter's basilica and the Vatican territory.

The European city: four approaches

It was noted in the introduction that writers such as Paasi (2001) had advocated a reading of Europe that stressed its significance as a social process. In many ways, the same argument applies to cities. Here, we have seen a range of work that could be called 'relational urban theory' (summarised well by Graham and Marvin 2001: 202–11). The likes of Manuel Castells (e.g. 1994, 1996, 1997a, b) have painted a picture of a (global) social world rewired and re(infra)structured around new communications technologies, with cities acting as nodes in the global economy. Such 'splintering urbanism' (Graham and Marvin 2001) is clearly adding to the social dislocation of the post-industrial economy, destroying the (perhaps mythical) coherence of the modern city. Simultaneously, there has been a growing interest in 'transnational urbanism' (e.g. Smith 2001) where flows of migrants are *fundamentally* reconstituting the urban cultures of a great many cities from Los Angeles (Smith 2001) to Stockholm or

Amsterdam (Hannerz 1996). As such, we should view the European city as being restructured by forces emanating from outside the old (metaphorical or not) city walls.

And yet there remains a powerful legacy of seeing these European cities as possessing some kind of 'essence' or innate character – not for nothing do many authors speak of the 'biography' of a city as if it were a person. Both in popular discourse and in branches of academia, the idea that cities possess sealed identities – even if influenced (as opposed to *structured*) by outside forces – is orthodox, a perfectly normal way of discussing these places. As I hinted above, it may be most fruitful to see these cities as constructed *by* such a tension.

To discuss this in more depth, I identify four literatures that might be used for structuring this idea of a Europe of the Cities: the entrepreneurial city, European Cities of Culture, the cityscape and palimpsest, and the city-state and neo-medievalism.

The entrepreneurial city: shifts in urban governance

My first cut at the European city starts with the growth in what Harvey (1989), has termed 'urban entrepreneurialism'. With its cities riven by deindustrialisation and govern-ments unwilling or unable to raise taxes to bail them out, urban managers such as mayors, chief executives, or head planners have come to treat the city as a corporation seeking to gain a market 'niche' in competition with other entrepreneurial city rivals. Thus urban policies – shaped by central government regulations – may seek to enhance the city's competitiveness, whether it be the offer to corporations of a large, low-wage workforce, or a well-trained, skilled workforce; for tourists; for 'command and control' functions such as corporate headquarters; and for government funding (Hall and Hubbard 1996, 1998; Swyngeduow *et al.* 2002). The battle for these are characterised by Harvey (1989) as a form of 'inter-urban competition'.

Across Europe, key areas of major cities are being redeveloped. Whether it is the incongruous office skyscrapers of the Centro Direzionale in Naples, the redevelopment of the Dublin waterfront, or Donau City in Vienna, there is a fairly radical transformation of the urban fabric (Swyngeduow *et al.* 2002). Yet despite their functional blandness, it has been suggested that at a time when globalisation is leading to a decline in sociability and communality, Europe's urban renaissance may be the best way of preserving democratic ideals and values:

> [M]ajor cities throughout Europe constitute the nervous system of both the economy and political system of the continent. The more national states fade in their role, the more cities emerge as a driving force in the making of the new European society ... [and so] we will be witnessing a constant struggle over the occupation of meaningful space in the main European cities, with business corporations trying to appropriate the beauty and tradition for their noble quarters, and urban countercultures making a stand on the use value of the city.

(Castells 1994: 23–5)

The competitive game gives a new twist to centuries-old rivalry (and co-operation) between cities, be it Paris–Lyon, Lisbon–Porto, Barcelona–Madrid, Edinburgh–Glasgow, or Moscow–St Petersburg. Furthermore, these can again be characterised as a battle between cities embodying a deep national 'essence' and those which are cosmopolitan, open to outside influence. In *All that is Solid Melts into Air,* Marshall Berman notes:

> the endlessly rich symbolism that developed around the polarity of Petersburg and Moscow: Petersburg representing all the foreign and cosmopolitan forces that flowed through Russian life, Moscow signifying all the accumulated indigenous and insular traditions of the Russian *narod*; Petersburg as the Enlightenment, and Moscow as anti-Enlightenment; Moscow as purity of blood and soil, Petersburg as pollution and miscegenation; Moscow as sacred, Petersburg as secular (or perhaps atheistic); Petersburg as Russia's head, Moscow as its heart.
>
> (Berman 1982: 175–6)

Thus despite the apparent novelty of urban entrepreneurialism, in many ways these forms of inter-urban competition overlay deeper narratives of civic rivalry. They are not purely economic battles. They may, as in the case of moving the capital of Germany from Bonn to Berlin, be implicated in deep debates over the geopolitical framing of the nation. Berlin, some argued, remained the city that had spawned the militaristic regimes of the Kaiser Reich, the Third Reich, and the German Democratic Republic. As Richie describes:

> When members of the 'Berlin as capital' campaign heard these arguments they were furious. How dare other Germans point to Berlin as the only place involved in Nazi crimes or in the Second World War? Were Bonn advocates claiming there were no Nazis in Bonn or Frankfurt or Bremen?
>
> (Richie 1998: 852)

By contrast, it was suggested by Berlin's defenders that the city was cosmopolitan and socially diverse, that it had become Germany's most politically literate city since 1945, and, indeed, that it had always opposed the Nazis electorally.

Cities of culture and urban tourism

> There has always been a close association between literature and cities. Here are the essential literary institutions: publishers, patrons, libraries, museums, bookshops, theatres, magazines. Here, too, are the intensities of cultural friction, and the frontiers of experience: the pressures, the novelties, the debates, the leisure, the money, the rapid change of personnel, the influx of visitors, the noise of many languages, the vivid trade in ideas and styles, the chance for artistic specialization.
>
> (Bradbury 1991: 96–7)

European cities are *old,* when compared with their counterparts in many other continents. From Rome and Athens (in antiquity), to Florence and Rome and Krakow (in the Renaissance), to Barcelona, Glasgow, and Vienna (in the modern industrial age)

European cities are renowned for developments in art, in architecture, in philosophy, in science. The sense of a culture of creativity or artistic innovation, as captured by Bradbury, has not always been valued in the modernist approaches to the city of the twentieth century, where universal solutions and technical fixes were prioritised.

In the context of post-industrialism, both capital cities and regional capitals (or 'second cities') have undertaken ambitious and wide-ranging cultural regeneration programmes. Furthermore, the European Commission has introduced the European City of Culture award, where every year a city from one member state would be showcased as a major arts centre. It was Glasgow's full-blooded use of their 1990 award that in many ways popularised what had hitherto been a rather elitist, or centralist, programme (Boyle and Hughes 1991). Prompted partly by the success of Glasgow, urban policy-makers have turned their attention to 'cultural regeneration' of old city cores, in a way that may be very different from the corporatised world of mainstream US urbanism. However, such an arts-based urban policy is still rooted in a realist reading of disinvestment and the need to regenerate ex-industrial cities. Thus as Evans (2001: 180) describes, 'the shift from arts-as-amenity to the cultural economy, whilst not seamless, has been justified as a pragmatic *and* ideological response to the decline in public cultural services and amenities . . . and at the same time to the disempowering effects of globalisation and commodification of much cultural production and consumption'. There are certain landmarks here: Glasgow's 'urban renaissance', Mitterrand's *grands projets*, Frankfurt's museum district, Barcelona's cultural quarter, the Bilbao Guggenheim, London's Tate Modern. So, it is clear that tourism as an industry is altering the fabric of Europe's built environment, as well as popular interpretations of its history. Judd's analysis of urban tourism is applicable more generally:

> For a tourist space to function, the image and the material product must complement each other; the city as a whole is made attractive to the tourist and accessible to the imagination through processes of *reduction* and *simplification*. The positive images projected by civic boosters and the advertising firms they hire amount to a coaching process: advertisements and tourist articles (such as those found in trade journals and airline magazines) interpret a city's essence, its history and culture, and tell the tourist what to do, even what to feel. Tourist images invariably invoke a romanticized, nostalgic sense of history and culture.
>
> (Judd 1999: 37, my emphasis)

This idea – of condensing the essences of place into easily communicated marketing processes – is part of what Judd calls the 'tourist bubble', the use of policing, urban design, and facilities such as restaurants and car parking to make the visiting process as standardised and risk-free as possible. This, taken to its logical extent by the Disney corporation, seeks total control of visitors and minimises any degree of unpredictability as a threat to profits. The Disney approach 'is the template of a privatized, consumption-oriented theme park intended to simulate a shared middle-class culture, aestheticise social differences, and offer a reassuring environment without arms, alcohol, drugs or homeless bums.' (GUST 1999: 100). However, it is not clear that this extreme approach to tourist marketing has become dominant in Europe. The theme park only accounts for a small percentage of tourist spending, and it could be argued that the relative and 'casual' safety of European city centres such as Brussels, Amsterdam, and Paris provides more of a draw to tourists than heavily designed environments.

Nonetheless, even if one accepts that European cultural tourism is of a 'higher' value than the theme park, all of this cultural regeneration may be missing the point. Jones (2000), points to the irony of Franz Kafka being hailed as a intellectual celebrity as part of Prague's attempt to cater to cultural tourism, even though his stay in the city was dominated by existential crises. The idea that art flourishes in adversity, as a critique of existing society, is developed in Dave Haslam's explanation of Manchester's leadership in alternative popular music production, a city that has gathered a reputation for fomenting some of the most significant British bands of the 1980s and 1990s, including Oasis, the Stone Roses, and New Order.

> Culture soon dies if it's confined to one city or one country. Artists of all kinds need to leap intellectual or emotional border controls . . . While our political leaders prevaricated over the propriety of the country's overseas alliances during the 1980s, the direction youth culture was taking was progressively more open-minded, international, cross-border . . . So a culture is created that's not about museums and national purity; it's a living culture which absorbs new things, remains open to new, old or foreign influence, and to new technology . . . This street culture is not about civic pride and marketing intiatives. It's about ideas, self-expression and escape. It can, and will, thrive in crappy coffee bars, under leaking roofs or down unlit roads.
>
> (Haslam 1999: 259–60)

In essence, Haslam suggests that adversity may spur on creativity. Interestingly, he highlights the fact that this music scene was inspired as much by American black musical influences as by anything indigenous, which highlights the significance of the cultural flows necessary for artistic creativity.

Cityscapes and palimpsest

In his book *Cityscapes of Modernity*, David Frisby identifies the concept of the 'cityscape', by which he means the problem of *representing* the city.

> Beyond the call to depict the modern city and our experiences of it that is shared by Baudelaire, the Impressionists, the Expressionists, the Surrealists and so on, there lay the problem of identifying the mode of representation itself. The 'panorama' of the city differs from the 'snapshot', just as the 'narrative' differs from the 'image'.
>
> (Frisby 2001)

The issue of how the city can be 'known' and *described* (written down) has been a recurring theme of many of the works cited in this book, from Pred's (1995) montage of Stockholm to Yalouri's (2001) fascinating account of the representational use of the Acropolis, to Maspero's (1994) argument for ignoring the centre city in favour of deep peripheral exploration (for more on this type of approach, see Pile and Thrift 2000).

Allied to this, we can consider the concept of the 'palimpsest'. The term derives from the earliest forms of printing, when wax tablets were used to carve images, then smoothed off and reused without completely erasing traces of the previous image. This idea can be applied to the urban landscape, where various fashions of architectural and planning theory and practice (from cathedral-building to industrial architecture to modernist town-planning to the conservationist movement) leave their traces, higgledy-piggledy, on the urban form. Visiting a city such as Naples, for example, is to visit a city cluttered with barely visible Roman ruins, narrow eighteenth-century streets, courtyards and royal squares, late nineteenth-century commercial galleries, hemmed-in cathedrals and hillside mansions, modernist office blocks and industrial port buildings and cranes, and – now – brightly coloured (postmodern) metro stations. These buildings represent *layers* of styles and space types, and these architectural treasures and monstrosities are being variously dusted down, tarted up, demolished or sand-blasted, and above all *preserved* due to the impetus they give to cultural independence and place marketing.

> Some of these big provincial cities, built upon commerce, feel like City-States still – more so, some of them, now they are fenced in by ring-roads, as by city walls ... Venice, for a thousand years a sovereign Power, still feels unmistakably like a City-State, and in Croatia the little city of Dubrovnik, ex-Ragusa ... bears itself within the glorious circuit of its walls as though it still rules its own fortunes. Milan is almost its own capital. Barcelona hardly feels subject to the Kingdom of Spain. Hamburg retained its liberal Hanseatic outlook throughout the Nazi period. Antwerp is the most truly polyglot city in Europe ... Bruges,

Figure 12 *Cathedrals, in this case in Palma de Mallorca, are a fundamental part of the medieval imagery that surrounds many European cities. Donald McNeill.*

> too, seems to me more consequential than Brussels. Riga in Latvia, seen from across its river, wonderfully suggests a City-State in an old engraving – castle, cathedral, spires and towers laid out in proud esplanade along the quays. Lyon and Marseilles are far more than just French provincial cities, and the rugged old Scottish city of Glasgow, once the second city of the British Empire, always feels as though it ought to have its own Government, conducting its own foreign relations and flying its own flag.
>
> (Morris 1998: 278)

Furthermore, Morris draws our attention to another key point – that the city walls of many European cities, so long the bastion against the semi-lawless patchwork of medieval Europe, have been replaced by their modern equivalent, the city ring-road. This, paradoxically, helps retain the shape of the traditional city – the city *intra muros* (within the walls).

City-states and neo-medievalism

Europe is given a distinct identity because of its *urbanity* and urban longevity, and it is argued that a Europe of the Cities may be a progressive vision of a future continent because of the mutual understandings fomented by urban life. Here, some commentators have suggested that cities may be superseding nation-states as a focus of loyalty. Pasqual Maragall, the long-serving mayor of Barcelona, has been particularly active in promoting this idea, lobbying hard to get cities recognised in the EU's Committee of the Regions.

Some commentators have been attracted to cities (and regions) within what has been called a new or neo-medievalism, where political sovereignty is not located in one place, but in several.

> We might imagine, for example, that the government of the United Kingdom had to share its authority on the one hand with authorities in Scotland, Wales, Wessex and elsewhere, and on the other hand with a European authority in Brussels and world authorities in New York and Geneva, to such an extent that the notion of its supremacy over the territory and people of the United Kingdom had no force ... We might imagine that the political loyalties of the inhabitants of, say, Glasgow, were so uncertain as between the authorities in Edinburgh, London, Brussels and New York that the government of the United Kingdom could not be assumed to enjoy any kind of primacy over the others, such as it possesses now. If such a state prevailed over the globe, this is what we may call, for want of a better term, a neo-medieval order.
>
> (Bull 1977: 255)

This idea of Europe's cities rediscovering some form of political autonomy is far-fetched. Most city councils have relatively tiny budgets compared with regional or national states, most lack any significant degree of political decision-making, but many possess and inspire a substantial amount of cultural identity. What is most interesting for our purposes is that, first, Europe may be the closest to such a neo-medieval form partly because nation-states have never fully eradicated pre-modern traditions, loyalties, and landscapes and, secondly, that since Bull was writing, the growth in cyberspace and globalising tendencies has further confused issues of territory and identity. By extension, there is the claim that cities are the best vantage point to view the emerging economic trends of a new Europe built on a knowledge-based economy. And, furthermore, some parts of cities – the City of London (Roberts and Kynaston 2001), or the Vatican (McNeill 2003), particularly – have an influence worldwide over cities that completely belies their territorial size.

However, while cities may not be sources of political sovereignty, they may indeed be important sites of collective and individual *identity*, as individuals make and remake neighbourhoods. Yet the dynamic nature of urban populations may in turn lead to feelings of *dispossession*, as I now discuss.

The city and the street

As noted in the introduction, many social scientists have argued for the importance of a general urban theory that allows us to compare and contrast the experience of very differing cities. By contrast, or perhaps in parallel, it makes perfect sense for many of us to talk about cities in an emotional way – we might defend our native cities fiercely in conversations, we might support its football team on certain occasions even if we hate football, we might moan with pleasure about the charms of some cities ('Oh, Paris is so romantic'), we might make the most appalling stereotypes of cities and their dwellers with complete earnestness ('Londoners are so unfriendly'). In this way, we are contributing to layers and layers of stories about cities that animate them, judge them, criticise them, or honour them.

Glaswegians, *Madrileños*, or Berliners have a peculiar 'personality'. This assertion could be justified on the grounds that, historically – perhaps through shared experience in times of crisis such as warfare or revolution or siege or famine or industrial class consciousness – there has built up a certain way of behaving which is either perceived by outsiders as distinctive, or is 'assumed' by its inhabitants to be an appropriate means of behaving. In this sense, cities and their denizens possess a 'native' way of life in the sense once used by anthropologists. As with most myths, they have a grain of truth, yet what is interesting is how they are fitted into major discourses about metropolitan life or urban change, be it place-marketing, professional humour (in films, television, newspapers, cabaret), or in the degree to which city 'natives' present themselves in public. An example of such 'civic identity' could be found in the Berlin council's project of stationing life-size bears (the city's emblem) at various points around the city, each caught in different poses or with different, often punning, messages.

The location of these ideal types of citizen is unclear, but they are often captured in the idea of 'street life', with the street being an important *scale* in European cultural, social, and political landscapes. This has two meanings. In the first, the street represents the sphere of action, of everyday life, which may suddenly be shattered when the impact of more abstract, elite political workings are stretched too far, or implode, or are challenged. It is thus more of a metaphor, of a representation. In the second, the street – taken to include all aspects of public space, such as metro systems, squares, and daily movement –

Figure 13 *One of Berlin's bears. Donald McNeill.*

is an actual site of urban experience. As such, it is a scale open to experience, emotion, psychological states, a place of gratification or fear or guilt or happiness.

However, an important theme in European urban criticism has been the idea of the death of the street (Fyfe 1998). The incursions of the motor car have served to undermine the street as a place of rich interaction; the dilution or removal of the denizens of the inner-city by demolition and/or gentrification has served to dull the vibrancy of the city; and the growing prevalence of privatised spaces and camera surveillance has apparently 'sanitised' some of these spaces. For some critics, this has been met with anger, a sense of cultural disorientation as familiar sites, people, and buildings are removed, or a politicised sense of injustice. Yet the street remains a complex terrain, and the renewed attention to urban design and anti-car measures, along with a renewed sense of the street as a forum for political protest, requires a careful consideration of its place within European society.

Pedestrians and car-drivers

The street has long had a cherished place in European culture, yet one which has been transformed drastically over the last 50 or so years. Rebecca Solnit's *Wanderlust: A History Of Walking*, describes this change as follows:

> Walking still covers the ground between cars and buildings and the short distances within the latter, but walking as a cultural activity, as a pleasure, as travel, as a way of getting around, is *fading*, and with it goes and ancient and profound relationship between body, world, and imagination.
>
> (Solnit 2000: 250, emphasis added)

Solnit is particularly concerned about the impact of the private car and the growth of the suburb, and while she may be accused of harbouring excessive nostalgia for bipedalism, her eloquent polemic captures – from an American perspective – precisely just why the European street is held in such regard. Yet in many ways, the idea of the casual city stroller – the *flâneur* – is being mixed in with a more aggressive, speeded-up city life, perhaps leading to a *decline* or erosion of street culture. This is happening in three main ways: first, the pedestrian is being replaced by faster ways of travelling (car, usually); second, this is also eroding the pattern of 'human scale' which cities like Paris and Barcelona are famous for, and which recent urban design innovations have sought to restore; third, this is often wrapped up in a corporate re-creation of heritage spaces, which replaces the messy practicalities of the vehicle-choked working city with a themed, pedestrianised version of urban life.

The classic conception of European street life has been threatened by several factors. Motorisation is one, given the heavy impact of car use on medieval city centres, and the paradoxical emptying or de-densification of the core as accessibility to the suburbs has improved. The pressures on those who have become economically marginalised is another, as rising rents have driven out those unable to retain their position in the heart of the city, or maintain the overheads of a small shop in the face of increasingly intensive retailing. And the perceived rise in crime has further helped big businesses in justifying processes of surveillance or exclusion, through the use of private security firms.

Second, cars and roads have fragmented cities, particularly during the phase of high modernism when environmental issues were often ignored. High speed roadways such as motorways and dual carriageways 'with highly limited access points have been superimposed on more fine-grained street structures' (Graham and Marvin 2001: 120), thus radically altering urban form, and in some cases carving apart existing communities. As Cobb (1998: 174) writes of Le Corbusier's visions for Paris:

> How he *hates* the place! With what sovereign contempt does he treat its mindless inhabitants! Here he is already in 1925 obsessed with his *grande croisée*, a swath of huge expressways cutting through the centre of the city, east-west and north-south, marking Paris like a hot-cross bun. And here he is again back in the 1920s, with plans to uproot

> Les Halles and most of the old street system between the rue de Richelieu and the rue Saint-Martin, and to erect on the vast quadrilateral of the old Right Bank thus devastated a series of tower blocks, the first of many versions of his alarmingly named *Ville Radieuse*.
>
> (Cobb 1998: 174)

Contemporary urban design and planning has, by and large, learnt from the failures of some of the more destructive tendencies of modernism. Increasingly, planners speak of a 'European' model of urban design and policy that carefully integrates various modes of public transport, pedestrianises streets and bans cars from congested city centres, beautifies existing public squares and spaces, and encourages buskers and performance artists, pavement cafes, and – in many cases – supports communities of urban dwellers. However, such novelties are not always met with universal acclaim, as I now describe.

Nostalgia, disorientation, and political anger

> Being Parisians, they themselves had for years watched their bustling *quartiers* being slowly transformed into museum-style shop windows ... they had watched the departure of an entire class of craftsmen, workers and small shopkeepers – all the people who went to make up a Paris street. They themselves had hung on, but saw renovation force out the poorly off, old people, and young couples with their children, who all disappeared as rents rose and flats were sold. Where did they all go? To the outskirts, to the suburbs. Paris had become a business hypermarket and a cultural Disneyland. Where had the life gone? To the suburbs. 'All around' could not, therefore, be a wasteland, but a land full of people and life. Real people and real life ... And if Paris had emptied, if it was no more than a ghost town, didn't that mean the true centre was now 'all around'?
>
> (Maspero 1994: 16)

The powerful argument presented by Maspero and explored in greater depth in the final chapter is that the experience of modern life based around the densely-populated central city has been diluted by a dual process of suburbanisation and gentrification. Maspero's argument fits with those of writers such as Chevalier (1994) and Cobb (1998) on Paris, Spring (1990) on Glasgow, or Vázquez Montalbán (1992a, b) on Barcelona. As these authors have eloquently captured, urban change can be a disorientating, anger-inspiring, or debilitating process for long-time residents of a city. The process of market-led gentrification, or the relocation of working-class people by comprehensive redevelopment of poor housing, has a very significant impact on urban identity. This can be thought of in three ways:

(1) As *nostalgia*, a sentimental yearning for the past ways of a city. There has always been a strong 'myth of community' in urban studies, which sees cities as organised *territorially*, rather than through distanciated friendships or networks. Ian Spring's *Phantom Village* (1990) examines the myth of the New Glasgow, as he calls it, an event centred on the 1990 European City of Culture award. In his extremely

insightful analysis of various media representations of the city, Spring argues that 'the New Glasgow is a city of rampant consumerism' (1990: 43), evident in a whole range of cultural artefacts from coffee-table books, cultural festivals, adverts, and films. His examples include the rediscovery of the native art nouveau architecture of Charles Rennie Mackintosh:

Old Mackintosh designs for furniture and street lampposts have been unearthed and there are plans to build from scratch a complete Mackintosh design ... House for an Art Lover. In city shops, art nouveau designs by Mackintosh ... adorn postcards, fingerplates, teatowels, designer mirrors etc ... Is the art nouveau revival triggered by a genuine interest in design or is it simply a potage of fashionable clichés, commercialised by the nostalgia boom? The importance is in the way that New Glasgow style repossesses the past to constitute an artificial indigenous tradition.

(Spring, 1990: 43–4)

The 'Tackintosh' Glasgow of trinkets and souvenirs was a significant precursor of the treatment of another architect practising a form of art nouveau, Gaudí in Barcelona. In 2002, local authorities declared the 'year of Gaudí', the 150th anniversary of his birth. While crowds flocked to the myriad of exhibitions, talks, and floodlit buildings, there were similar feelings that, as in Glasgow, the city was being turned into a kind of architectural theme park. In Barcelona the complex Catalan nationalist and religious imagery that permeates Gaudí's work was being commercialised and decontextualised.

(2) As *disorientation*, the drastic redevelopment of old city centres, the opening of new bars and restaurants, the spread of 'gay villages' such as Soho in London or Chueca in Madrid, the presence of young urban professionals, often derided as materialistic 'yuppies' means that the working class cultural institutions of central cities are replaced. In Paris, Cobb notes with sadness the disappearance of the street life of the old left and right bank *arrondissements* (districts):

The rue Mouffetard (or what has been left of it), once the most popular market of the Left Bank, and a street of intense sociability, has been given over to tourism, bars, and pornography. Only a few old inn-signs ... hang as sad, limp reminder of what had once been a triumphal gallery of small shops, barrows, overhanging clothes-lines, and popular eloquence. *Les gens de la Mouffe* have been spirited away, as on a magic carpet, to Sarcelles or to another of the New Towns of the Paris region ... Their place has been taken by young *pied-noir* couples running pizza bars and by elegant antique-dealers.

(Cobb 1998: 180–1)

Furthermore, the cultural and social disorientation may be increased by the *physical* changes to neighbourhoods, with the demolition of old buildings and landmarks actually making the landscape more difficult to read even for long-time residents of a city.

(3) As *political anger*, the growth in entrepreneurial modes of urban governance (described above) has often been associated with a political climate increasingly dominated by business, rather than social, values. Many former left-wing politicians

are now abandoning their traditional working class base and are targeting policies at a middle class audience. In the writings of Manuel Vázquez Montalbán on Barcelona, the changing nature of the city's old working class neighbourhoods such as the Barrio Chino or Poble Nou during and after the 1992 Olympics was a politically charged process. This process, explored in his detective novels (e.g. Vázquez Montalbán 1990), is represented in the property deals undertaken by the social democratic council, and a gradual process of gentrification that has worked to undermine the traditional, low-waged inhabitants of the centre (see also McNeill 1999b, Chapter 2 on Vázquez Montalbán's Barcelona, and Haslam's searing 1999 critique of the 'new Manchester').

The link between each of these responses and manners of critique is their *emotional* nature, which means that as interpretations, they are partial and political.

Figure 14 *Demolition in central Berlin. Donald McNeill.*

Box 4a   Barcelona's La Rambla

Barcelona is famous for its tree-lined boulevard known as La Rambla. This beautiful street, hemmed in by tall apartments, shaded by a long canopy of lush plane trees, animated by shops, bars, and pavement cafes, epitomises how enjoyable a pedestrian-centred city culture can be. Buskers, human statues, newspaper kiosks, and flower stalls slow movement to a gentle stroll. Cars are limited to a single-lane of traffic on either side of the broad central walkway (a common feature of many Spanish towns and cities). However, some critics see the Rambla's charms dissipating, a victim of its popularity among tourists, as more of a theme park than a 'civic' space (Espadaler 2002). Swollen with crowds, the street is increasingly log-jammed, its role as a barometer of social and political identity replaced, or at least hindered, by an array of performance artists (who stifle the few genuine artistes), the protrusion of metallic cafe chairs, caricaturists, and trinket stalls.

Ironically, therefore, the idea that 'walking the city' is a feature of citizenship, democracy and freedom is curtailed not by political decree, but by commercial success. La Rambla has an important place within the city's history, as much as a centre of its historical civic sense (it is the centre of the annual parades of giants, and other civic rituals), as in the political transformations that the city and nation(s) have undergone since the 1970s.

Here is Manuel Vázquez Montalbán's view of the Rambla of 1976 in his novel *The Angst-Ridden Executive*:

> As night settled on the Rambla, Carvalho began to register the symptoms that marked the onset of the daily confrontation. The riot squad had begun moving into position, according to the prescribed rituals of the ongoing state of siege. Apolitical counter-cultural youth and young counter-cultural politicos maintained their customary distance from each other. At any moment a gang of ultra-right provocateurs might appear, and you would see the militants of this or that party disperse and head for their now legalised party offices ... Between the hours of eight and ten the prostitutes, the pimps, the gays and the crooks great and small would disappear off the streets so as not to find themselves caught up in a political battle that was not of their making.
>
> (Vázquez Montalbán 1990: 85)

This view of the Rambla (and George Orwell (1938/1989) made its earlier political significance famous in *Homage to Catalonia*) holds up the condition of the street, and the behaviour of its crowds, as being symptomatic of the political and social state of the city. During the transition it became a somewhat traumatic red light area, with not infrequent cases of drug-fuelled physical attack (and even murder) serving to disconnect the old town from the richer, middle class districts of the city. Today, it remains a site of intense social and cultural mixing; whether the street can retain such a huge burden remains to be seen.

Politics from the streets

> Public marches mingle the language of the pilgrimage, in which one walks to demonstrate one's commitment, with the strike's picket line, in which one demonstrates the strength of one's group and one's persistence by pacing back and forth, and the festival, in which the boundaries between strangers recede. Walking becomes testifying. Many marches arrive at rally points, *but the rallies generally turn participants back into audiences for a few select speakers;* I myself have often been deeply moved by walking through the streets en masse and deeply bored by the events after arrival. Most parades and processions are commemorative, and this moving through the space of the city to commemorate other times knits together time and place, memory and possibility, city and citizen, into a vital whole, a ceremonial space in which history can be made ... Such walking is a bodily demonstration of political or cultural conviction and one of the most universally available forms of public expression. It could be called marching, in that it is common movement toward a common goal, but the participants *have not surrendered their individuality* as have those soldiers whose lockstep signifies that they have become interchangeable units under an absolute authority.
>
> (Solnit 2000: 216–17, emphasis added)

The early years of the new millennium have been punctuated by a rising tide of anger and political organisation not dissimilar to that seen in 1968. Furthermore, the targets of these demonstrations – against a growing culture of corporate manipulation of everyday life – have clear echoes with the complaints of the disaffected protesters of the late 1960s. The message – a strong international solidarity – is similar, and the place to demonstrate that solidarity is similar. It is the street.

Following the events of the anti-globalisation protests in Seattle in 1999, summit after summit of government leaders across Europe have been disrupted by a broad, largely non-ideological coalition of forces. Environmental pressure groups, trades unions, farmers, the unemployed, 'drop the debt' groups, ATTAC (which campaigns for the taxation of cross-border financial transactions), small-scale farmers, and anti-consumerist groups have united in cities across the continent to protest against the growing stranglehold of corporations on European life. In Genoa, in late July 2001, this upsurge in feeling culminated in one of the most controversial events to date in this growing movement – the wide-scale rioting and street battles that accompanied the G7 summit (the leaders of the world's richest countries) in the northern Italian city. The death of one protestor – Carlo Giuliani – from a police bullet shocked many. The subsequent beatings of demonstrators by a vengeful Italian police squad highlighted the fact that this was not a domestic affair: those injured were drawn from a range of European countries. The numbers on the event – reasonable estimates put the figure between 200,000 to 300,000 – showed the scale of organisation. And the drastic measures taken by police to try to protect the world leaders – caging off the central core of the city and sealing the harbour – was a metaphor for the 'democratic' deficit that prevails in contemporary political life.

The recriminations over the battle for Genoa rumble on. Many peaceful groups felt that the message of the protest, namely the exploitation of third world labourers and resources by Western corporations was lost in the bloodbath. Others pointed to the uncontrolled anarchist groupings allowed to assemble in the city, with accusations that they were fuelled by police insiders. Whether or not this was the case, few would argue that this new movement, with the target of its anger not the state but the corporate board-room, has reassumed the mantle of the street demonstrators of the late 1960s.

Yet while the events of Genoa were compared with Paris in 1968, the cities of East and Central Europe know street protest in great depth. In 1989, the likes of Leipzig, Prague, and Bucharest witnessed scenes of great emotion and, in the latter case, great brutality, as people took to the streets to oust the communist governments. The tactics used in these protests were often very creative, as demonstrations showing opposition to Serbian leader Slobodan Milosevic showed in the mid-1990s. Milosevic had repressed any attempt to challenge his interventions in the Balkan war, but by 1996 the only way he and his party could retain power was through electoral fraud, a monopoly on the mass media, and the use of police and the military to control the streets. A diverse opposition movement coalesced from a variety of sources and, as Jansen (2001) records, organised unusual and imaginative forms of protest. For example, at 7.30 each evening, when the main state news report would be broadcast, protestors would use a medley of instruments from pots and pans to fireworks to 'drown out' the censored news programme, with pirate radio stations acting as a alternative news channel.

> Reclaiming control over Beograd did not only involve the politicised insertion of human bodies into public space. Another central element was noise: shouting, singing, music from sound systems, and so on. Usually, the demonstrations took place in the afternoon, but certain areas were filled with noise, and especially with the piercing sound of whistles, virtually day and night. More than anything the whistles became the emblem of these protests: people wore them conspicuously, they appeared on posters, stickers and postcards. Blowing a whistle and filling the air with noise was an integral part of the metaphoric process whereby the urban space of Beograd was to represent the field of politics. This had the simultaneous effects of territorialisation, through the imposition of noise upon urban space, and deterritorialisation, through the ungraspable nature of a noise invasion into regime-controlled space.
>
> (Jansen 2001: 40)

The future of the central urban street as a site of protest remains open to debate, but in the 35 years between 1968 and 2003, it has remained a very significant forum of political expression.

Thus the European street – as the site of public interaction, civic identity, movement and circulation, political protest – is an important site of both representational and bodily practice. In the changing cityscapes of the New Europe, it is returning to a central role.

Transnationalism and the European city

Transnationalism has arrived as a major field in contemporary studies of globalisation (Smith 2001; Vertovec 2001). In many ways this is due to the undeniable rise of flows of people between countries for a whole variety of reasons. Either fleeing political persecution (e.g. Kurds), warfare (e.g. Vietnamese), or intense poverty (e.g. Albanians), the transnational impulse is one manifestation of the misleading nature of the traditional map of nation-states. The nature and regulation of migration within a 'fortress Europe' is one that I will pick up in Chapter 6, but here I want to consider the light it sheds on understanding the European city. The experience is of global flows and migrations, as well as European ones. For example, in London the 1991 census recorded 21,000 French, 32,000 Germans, 30,000 Italians, 19,000 Spanish, and 13,000 Portuguese citizens. Yet it also recorded 33,000 Americans, 23,000 Australians, 12,000 Canadians, and 17,000 Japanese. (White 2001: 141). And this says nothing of the many thousands of British citizens who migrated from the Caribbean, India, Bangladesh, Pakistan, and Kenya and Uganda over the preceding decades. The same kind of story – which echoes the colonial ties of the past – will be found in Lisbon, in Brussels, in Berlin, in Rome, in Paris.

To conceptualise cities as being based on flows rather than essences, on relation rather than fixity, is an important challenge. One of the foundational texts in this area is Doreen Massey's (1991) essay on Kilburn High Road, an area of predominantly working-class North West London. In this, she stresses the porosity of places, and the coexistence and mingling (either peacefully or with some tension) of different identities. Smith summarises her stroll down this vibrant street:

> In addition to the many signs of an Irish presence and IRA political activity, she gazes upon saris in Indian models in shop windows, chats with a Muslim about the Gulf War, watches airplanes pass overhead, and confronts a traffic jam of cars leaving London. This simple exercise in participant-observation ethnography is a useful way to 'map' places without drawing fixed boundaries around them.
>
> (Smith 2001: 107)

Massey has been criticised for this essay, not least because she appears to assume that Kilburn High Road is a kind of 'anyplace', whereas it is probably one of the most multi-ethnic parts of Britain. Yet Smith is right in drawing attention to the potential of her method. As with Maspero's 1994 tour of the Paris suburbs (discussed in Chapter 7), even the most basic forms of travel writing that are written from a standpoint of transnationalism (as opposed to the usual stance of essentialism) can expand our geographical knowledges. In this context, I want to make five main points.

First, the use of ethnographic method has become of paramount importance in sociological and geographical research into European cities. Hannerz (1996) provides two fine-grained essays on Amsterdam and Stockholm, in which he reflects on the porosity of each city's transnational relationships (see also Pred 2000). In such work, the positive forms of cultural mixing and hybridisation of European societies are often seized upon by academics who carry out more or less sensitive, qualitative studies using methods such as life histories, participant observation and semi-structured interviews. Of course, there are dangers in pursuing these 'hidden' populations uncritically. As Smith notes (2001: 139), there is a 'temptation to capture the essence of a "local voice," inscribing it as a heroic individual or collective challenge to the oppressive forces of global modernity', a tendency to see the migrant experience as marginalised, victimised, and exoticised. Nonetheless, the complexity of transnational identities and mobilities is such that only careful ethnographic approaches can provide a full picture of, for example, transnational business practices and multiple loyalties.

Second, these cities have always been strongly shaped by migration flows and imperial, colonial and post-colonial linkages, even in the cities in the smaller member states of the EU:

> One can perhaps more readily take the Pakistanis and Jamaicans in London and the Senegalese in Paris for granted, since London and Paris are undisputedly world cities of the first rank. But the Surinamese and the Javanese in Amsterdam, whether first, second, or third generation, are more like the Angolans and the Goans in Lisbon . . . Colonial and postcolonial memories are probably never unambiguously happy, on one side or the other, but in different ways old metropoles, *even in what have become small countries,* remain centers to which scholars, tourists, and sometimes exiles from old dependencies continue to be drawn.
>
> (Hannerz 1996: 145, emphasis added)

Third, unlike traditional anthropological work which focused on a tightly defined community, territorially bounded groups, it is clear that these diasporic communities are mobile and connected, making for fascinating trans-global cultural flows. It is therefore misleading to think of transnationals as having *one* home. Their lifeworlds are split. Smith emphasises the complexities of the lives of migrants, 'who are currently seeking to orchestrate meaningful lives under conditions in which their life-worlds are neither "here" nor "there" but at once *both* "here" and "there"' (Smith 2001: 151). Clearly, the ability of these migrants to move is constrained by many things, be it money or the political situation in the country of their territorial sense of belonging. Furthermore, migrants' earnings in the metropolitan centres of advanced capitalism are likely to have profound effects on their' lands of origin, both positively – in the building of new health and education facilities, for example – and negatively, in that it may disrupt local community structures, labour markets, and fuel prices (Vertovec 2001: 575).

Fourth, ethnic minorities are likely to be subject to some form of ghettoisation in European cities, although as Çağlar (2001) shows in relation to Turkish youth in Berlin, this ethnic stereotyping may mask the increasingly cosmopolitan identities adopted by young people engaged in transnational networks. Furthermore, Andall's (2002) study of second generation African-Italians in Milan suggests that many of her respondents view further European integration as involving an improvement in their position, that 'Europeanisation' may overcome the 'mutually exclusive' perception of being black and being Italian.

Fifth, ethnic businesses are very significant components of the urban economy in major cities. While these are often stereotyped as being dominated by the fast-food and sweatshop modes of production, recent work has demonstrated the complexity of a lot of ethnic business, both in terms of product design and locational decision. Leung (2001) discusses the business practices of Taiwanese computer businesses in Hamburg, high-lighting the intersection between a whole range of factors (rather than ethnicity pure and simple) that constitute the activities of groups there. Nonetheless, these entrepreneurs are able to 'organise their linkages locally and transnationally, mostly through ethnic networks', which includes trading on the global competitiveness of the Taiwanese computer manufacturing industry, allowing them a flexibility within Europe's volatile computer retailing market.

The field of transnationalism is a complex one, and the term can be prone to overuse. As Vertovec (2001: 576–7) notes, it may occlude historical forms of migration, it may be used to describe too broad a range of phenomena, and it may be subject to a techno-logical determinism which implies that the development of satellite technology and cheaper air over-rides human motivations for migration. Nonetheless, in the context of the European city it underscores the porosity the city, the flexibility of cultural networks, and the cosmopolitanism that can undermine 'rooted' national identities.

Mayors and city politics

The theme of this chapter has been the idea that the European city is both fixed and networked, defined both by bounded place histories and 'essences', and also by its relation to other places. Here, I want to argue that this tension is often carried through in the activities of the city *mayor*. Interestingly, these politicians are often as identifiable as government ministers or regional presidents, yet they usually lack any significant budget and are largely lacking in political power. However, they can substitute for this by manipulating the cultural weight of their cities. During the 1990s, a constellation of factors contributed to the high media profile of mayors, as follows (from McNeill 2001b):

- The reestablishment of a mayor for Paris in the 1970s for the first time in almost 100 years, which Jacques Chirac used to contest the Mitterrand presidency and ultimately 'jump scales' to become French president (Collard 1996); and the more general idea of *cumul des mandats*, whereby French political figures in prominent posts are often concurrently mayors in their local territories.

- The global profile enjoyed by Pasqual Maragall by virtue of the successful staging of the 1992 Barcelona Olympics; (McNeill 1999b, 2001a).

- The emergence of a new class of Italian politician after the discrediting of the old corrupt 'partitocrazia' in the early 1990s, with mayors such as Cacciari (Venice), Rutelli (Rome), Orlando (Palermo), and Bassolino (Naples) gaining high national profile by virtue of their city leadership. All acted as ministers at some point during their period in office, with Rutelli emerging as the leader of the Italian Left in the 2001 general election.

- The interest in Rudy Giuliani's 'zero tolerance' approach to crime in New York, a policy which many European mayors considered as a means of 'cleaning up' their cities. Furthermore, these ideas were often incorporated into *national* crime policy.

- The creation of a directly elected mayor of London by the Labour administration, (partly inspired by the Giuliani executive style of leadership) which was seen as a means of cutting local government bureaucracy. The controversy surrounding Ken Livingstone's non-selection and his subsequent victory as an independent candidate gave the post and the 2000 election a huge degree of publicity worldwide, framed above all in terms of the personality clash between Livingstone and prime minister Tony Blair.

- The creation of new fora for mayoral collaboration, particularly the Eurocities lobbying movement, the EU's Committee of the Regions, and the Council for European Municipalities and Regions.

- The election of openly gay mayors in both Berlin and Paris in the late 1990s, the latter of whom was stabbed in a (failed) assassination attempt.

So, for a variety of reasons, mayors have emerged as significant political players beyond the immediate confines of their cities. Despite the legal and resource constraints placed on urban leadership, mayors – especially when directly elected – can claim a democratic mandate that many prime ministers do not share. Along with the expectation that they are able to run and provide urban services efficiently, they are able to move into the more ideological sphere of culture, citizenship rights, crime, and many other issues. In certain circumstances, they are able to transcend their mandates and play a significant role, be this in regional or national politics, or indeed in diplomatic roles.

(1) Embodying the city

European mayors have in many ways sought to actively embody many of the stereotypical attributes of their city, and may use this to link a local dependence on a particular place to holding office on a national scale. In France, for example, the *cumul des mandats* – where politicians may simultaneously be city mayors, members of parliament, and government ministers – means that prominent politicians may actively bridge the gap between region or locality and capital city. Mayors, or urban leaders, thus *embody* their city particularly in terms of representing it in a wider cultural and political space.

The idea of *lineage* is also important, that mayors can effectively follow in the footsteps of illustrious predecessors. Francesco Rutelli in Rome initially attempted to inherit the garb of early twentieth-century mayor Ernest Nathan (McNeill 2001c, 2003). Yet the place in popular mythology that these mayors enjoy is a concrete fact in explaining governance capacity, through their ability to influence both extra-local actors (either in lobbying for public spending in return for vote delivery, or through attracting external private capital) and locally dependent actors, be they party colleagues, or the embedded private sector.

(2) Performing elections

Electoral campaigning (which, after all, can be seen to be a persistent factor in a politician's entire period in office) often uses the city as a kind of political theatre, and as the mediatisation of politics has become more sophisticated, so symbolic acts, gestures, and personality 'performances' become more important. We might thus see elections as condensations of four or five-year cycles of governance, which are imbued with narratives of place identity, contested accounts of policy and economic performance, and so on. So what is the role of 'local' media in constructing the themes and discursive boundaries of the campaign, given the significance – particularly in capital cities – of newspapers such as the *Evening Standard* (London), *Il Messaggero* (Rome) or *La Vanguardia* (Barcelona) in shaping urban political opinion?

(3) Cityscape as theatre

This may be the commissioning or unveiling of megaprojects such as opera houses or urban spectacle, such as Maragall with the 1992 Olympics, or Bassolino with the G7 summit in Naples. Furthermore, these are likely to intensify latent city-region or city-nation state tensions if there is a clear conflict over symbolic politics, such as between Maragall and Catalan president Pujol in 1992 (McNeill 1999b). The power of such events in 'branding' or legitimising a broader hegemonic project has been explored by, for example, Kearns (1993), who documents the conflict between Chirac and Mitterrand over the staging of the 1989 bicentennial celebrations in Paris. Yet smaller scale 'pavement politics' is also important, from Chirac's commitment to municipal street-cleaning in Paris, to Bassolino's rescuing of Naples' Piazza Plebiscito from the automobile, to Maragall's commitment to public space in Barcelona. It is worth reflecting on the use of these landscapes as backcloths to mayoral reign, again allowing leverage in attempts to jump scales through symbolic performance.

Thus it is important to consider how city-regions might be *shaped* in the broadest sense by urban leaders. This will include a raft of issues – architectural, legal, economic, spatial planning, policing and social policy, morphological, ethnic, which politically motivated leaders are likely to pursue through a short-term electoral rationality. How is reterritorialisation *embodied,* how is it *discursively or materially represented,* how is the gap in scales and levels of government negotiated? Who are the actors transgressing or collapsing or reinforcing spatial scales, who are the leaders and collectivities shaping this new agenda?

4b   The Vatican and global Catholicism

Thou art Peter, and upon this rock I will build my church: and the gates of hell shall not prevail against it. And I will give unto thee the keys of the kingdom of heaven: and whatsoever thou shalt bind on earth, shall be bound in heaven, and whatsoever thou shalt loose on earth shall be loosed in heaven.

(Matthew 16: 13–20, in Lees-Milne 1967: 26)

In 2000, the city of Rome celebrated a Holy Year, or Jubilee, a practice that the Roman Catholic Church had instigated in 1300. Hundreds of thousands of pilgrims visited the city, swelling tourist takings, filling churches, and reaffirming the role of Rome as a spiritual capital for millions of Catholics. The Jubilee was at the forefront of the Vatican's attempts to (re)place itself at the centre of contemporary processes of globalisation, reasserting Christian Catholic values at a time of intensifying secularisation. At first sight, therefore, the Jubilee 2000 reinserted the Vatican's claim to universal – spiritual – status. Yet the staging of the event in Rome required the support of the secular Italian state, particularly city and central government. This has involved a far-reaching examination of the mutual

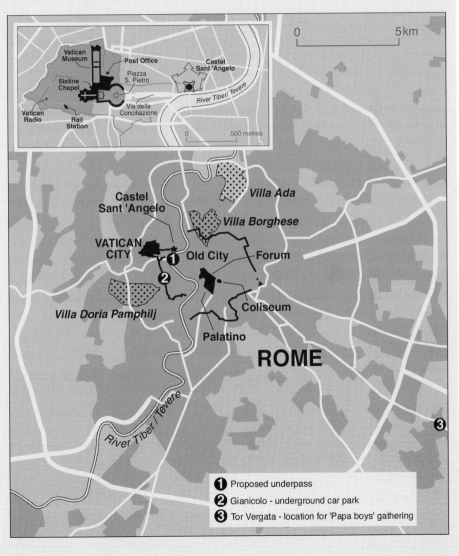

Figure 15 *Rome and the Vatican (map).*

rights and duties of Church and state in a context where the Vatican holds territorial sovereignty in a tiny island in the midst of the Italian capital, which led to a deep politicisation of the event. From gay rights activists to archaeologists, the impact of a huge religious event on the rights and relics of secular Italy was fiercely contested.

The legitimacy of Rome as centre of the Catholic Church today comes from the words that opened this section, interpreted by Lees-Milne as follows:

> Peter the rock is to be the foundation of the Christian Church. To him are given the keys which admit to or exclude from the kingdom of heaven, which kingdom, Christ says at another time, 'is not of this world'. Whomsoever and whatsoever Peter approves or condemns in this world shall be approved or condemned in the next. Can Christ therefore have possibly meant, the Catholic Church argues, that this tremendous authority and power should be granted to one middle-aged man to be exercised solely during his comparatively short mortal life? In the Church's opinion such a limited delegation would have little point or sense. Clearly therefore if Peter was to become Christ's delegate on earth, Peter must in due course be followed by a succession of others with the same authority until the day of judgement. And so in the eyes of Catholics it has come about. It is upon what she believes to be this divine revelation that the Catholic Church has built the towering fabric of the apostolic succession and the doctrine that the inheritance of St Peter's authority is of divine origin traceable from Pope Paul VI back without a break in the chain to Jesus Christ.
>
> (Lees-Milne 1967: 26)

The 'towering fabric' to which Lees-Milne refers is, of course, the basilica and square of St Peter's, which has become of growing importance to the papacy since the Reformation. This Petrine lineage is rejected by Protestant theorists, and thus this material, performance space is the home of the Pope, and represents the distinctive location of Roman Catholicism, the most influential religion on earth if based on number of followers.

St Peter's is, of course, the most famous building in the world's smallest independent state. With 108.7 acres, and approximately 500 residents, the Vatican (or Holy See) consists of the Papal residences but also the Curia, which includes the main theological offices of the Catholic religion. The Vatican museums – which include the Sistine chapel – are a major tourist attraction, with their priceless collection of art. Of course, the Vatican is primarily a major religious centre. During the Holy Year of 2000, Rome and the Vatican were visited by an estimated 46 million people, the vast majority of them Catholic pilgrims from around the world (and including the 'papa boys', a huge gathering of Catholic youth). Here, as Chidester has argued

> Traditionally, Christian pilgrims formed a communal relationship with each other and with the caretakers of the sacred sites throughout the ritual of pilgrimage. That communal relationship, which was forged through the hazards of the journey, the sharing of resources, and the entry into an alternative reality, defined the social character of the

Figure 16 *St Peter's in Rome, awaiting Papal mass. Donald McNeill.*

pilgrimage. In the modern tourism industry, however, the pilgrim's social role was defined as that of the client who paid for travel arrangements, airfare, tourist guides, comfortable accommodation, and pre-packaged experiences of the sacred.

(Chidester 2001: 593)

In many ways, then, the centrality of Rome in a medieval world of believers is as strong as it was, but has been transformed by the 'commodification' of religion that has occurred during modern times. Nonetheless, the Vatican acts as a magnet for the devout, and – with its visitors from Africa, the Phillipines, Latin America, the US, and Eastern Europe, among others is a clear example of how Europe's cities are transnational receivers of human flows from around the world.

Furthermore, it emits ideas from the mythically-justified, but materially constituted set of offices that form part of the Vatican. Here, it has been argued that the presence of the theologically conservative John Paul II and Curia (the Vatican 'civil service') serves to water down progressive policy in the Church, as happened after the Second Vatican Council in the 1960s. Dissenting bishops and theologians – such as the Brazilian Leonardo Boff, who advocated a left-wing liberation theology to address the plight of the poor in Latin America, were summoned to the Vatican for interview, and subsequently barred from preaching. To this end, the Vatican functions as a type of 'command and control' headquarters over the huge global institution of the Roman Catholic Church (McNeill 2003).

Disembedding cities: football and television

The relationship of the football club to the European city is extremely important. In the context of developing the everyday European consciousness discussed in Chapter 1, football often provides Europeans with a mental map of cities as diverse as Kiev, Turin, and Rotterdam. Clubs have often been flag carriers in embodying the 'essence' of cities and civic pride. However, the changing relationship between sport and place, particularly in a dramatically changing televisual market for sport, means that this relationship is diluting the role of the club as a vehicle for civic identity. For these reasons, then, football clubs are dramatically part of the idea of Europe as a space of flows, whether this be match coverage beamed around the world on television, supporters travelling abroad to matches, players being transferred between clubs, and so on. So they tell us something about *both* the nature of city essence and of relations *between* European cities.

First, there is the *changing crowd, and the changing stadium*. Football was a classic popular pursuit, booming across Europe throughout the twentieth century. The vast concrete bowls such as the Estadio da Luz in Lisbon, the municipal stadia of Italy, and the tightly-hemmed English grounds such as Arsenal's Highbury or Liverpool's Anfield both reflected the barely-regulated nature of mass audience participation. Yet with fears over crowd safety and a growing desire to attract a family audience, both the crowd and the stadium in Europe have been transformed. The all-seated, increasingly luxurious stadium has now appeared, with prices to match:

> With income from such audiences going directly to clubs and their media partners, such changes will truly herald the age of the 'electronic turnstile' and the electronic football 'crowd'; a shift which has already produced, so it is rumoured, from Milan's owner and media magnate Silvio Berlusconi the not altogether whimsical observation that soon football fans may be admitted to games free of charge as the necessary 'extras' for satisfying TV pictures of the sport.
>
> (Williams 1997: 250)

(a)

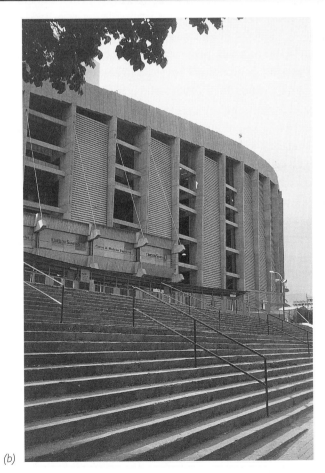

(b)

Figure 17a, b *Barcelona vs Real Madrid, Camp Nou football stadium. Donald McNeill.*

Growing corporate sophistication may mirror the situation in the US where league membership may be sold on the basis of franchises. The likes of Milan or Manchester United now have a value that may have little to do with the wishes of supporters resident in these cities (or indeed exiles).

Second, there is the *changing player*. A very clear impact of European Union regulation on the sport came with the so-called Bosman ruling, which forced through a freedom of movement between European countries. This was part of a trend towards an increasing number of 'cheque book' players, those who play for bonuses rather than their club's identity. Interestingly, football must be the purest market for cross-border labour flows in a single European market, as skills are not limited by linguistic barriers and 'market knowledge' is highly developed on an explicitly European stage. This is linked to the importance of branding: the likes of Nike and Adidas use star players in a portfolio that can be extended to trans-European advertising and the marketing of standard products – football boots, replica shirts – to a huge audience.

Third, there is an apparent *depoliticisation of football*. While I noted above that there has been a very real civic rivalry between teams from different cities (and often from the same city), it is clear that there has been a general penetration of market rationality that has now diluted the fierce social attachments that social groups had to their football clubs. The rivalry between Rangers and Celtic in Glasgow, or between Barcelona and Real Madrid in Spain is still a spectacle, but shorn of much of its symbolism. Similarly, in post-communist Europe some of the major teams of the post-war period, such as Dukla Prague or Dinamo Berlin, both of which were closely associated with their regimes, have virtually disappeared post-1989.

The case of FC Barcelona (Barça) warrants a brief discussion, as it is the Catalan club which has arguably changed its relationship to its city most markedly in the last decade or so. The change was associated very closely with the figure of José-Luis Nuñéz, elected president of the club in 1978. It was Nuñéz, a pro-monarchy property developer who effectively severed the club's ties with Johann Cruyff, hero of the anti-Francoist spirit of the 1970s (see Burns 1999, Chapters 1 and 17 for a full discussion). While the club and its successes on the field (especially against the regime-tainted Real Madrid) were among the main strands of anti-Francoist protest and opposition during the 1970s, by the late 1990s the team was largely composed of non-Catalans, and had a notoriously fickle support.

Fourth, there is the growing impact of the *media corporation* on the major football club, demonstrated in media magnate Rupert Murdoch's attempt to take control of Manchester United. One of the richest and most identifiable clubs in the world, United have transcended their city in terms of both support and playing staff. As such, they have been at the forefront of attempts to create a pan-European league, drawing on the lucrative opportunities offered by new pay-per-view television marketing, and the advertising and sponsorship and branding deals that follow on from it. In 1998, Rupert Murdoch's BSkyB bid to buy the club was headline news, drawing politicians – as well as the club's supporters – into one of the highest profile cases of monopoly regulation in the 1990s. The issue was clear: the global media mogul could use the club's brand profile and huge virtual audience to corner the satellite television market. Their regular presence in the Champions League allowed access to a European market and beyond. And the fame of many of their players also increased their attractiveness as a package that transcended their 'Manchester' location. As McGill (2001: 230) notes, the club's 'global reach' is remarkable: 'Research from China shows that United has an almost 80 per cent unprompted name awareness . . . This level of rating is phenomenal in marketing terms and can only be beaten by multinationals like Nike, McDonalds, Adidas . . . '. Ultimately, and to the surprise of many given the close relationship between Murdoch and Tony Blair, the UK government blocked the bid. Yet the trend had been set, and it remains to be seen whether football can escape the clutches of corporate raiders (see McGill 2001: 229–36).

So, football has been transformed in Europe in recent years, reflecting more general trends in which places are 'disembedded' from their pasts. Yet alongside this feeling of the dilution of the place specificity of football clubs, with their dispersed fan base and foreign 'cheque book' players, there is a sense in which it may be fomenting a greater European popular consciousness. Kuper's (2002) tongue-in-cheek essay on the positive effects on the UK of an 'invasion' of stars from Eric Cantona to Ruud Gullit is that 'Europe' is a more positive cultural model than Britain, from diet to dress sense to level of education and linguistic ability.

Conclusion

There are four issues I want to bring out of this chapter. First, thinking on European cities is dominated by a focus on the centre, often the location of the city's public institutions (city hall, opera house, museum, nineteenth-century (or older) shopping arcades, and so on). While suffering from urban decline during the post-war period, they are now undergoing a renaissance in property values, design, and attractiveness for tourists. They are often seen as the location of 'culture', usually in the sense of high-brow art, music, and architecture. Second, these cities are very often discussed in a popular way as having certain 'essences', a set of geographical and social attributes which hold substantial mythical force. People make reference to 'Berliners', 'Mancunians', 'Milanesi' as possessing certain psychologies and character traits. While social scientists often shun such essences, they are nonetheless central to the public discourse of cities, and many residents may actively perform such identities, especially to outsiders.

Third, by contrast, insights from 'transnational urbanism' suggest that these unchanging essences may be replaced or transformed given that these cities have significant minority ethnic populations that reflect an imperial past and a post-colonial situation of migration. Here, it is almost impossible to sustain the idea of 'insiders' and 'locals'. Fourth, under an increasing globalisation of capital flows and investment decisions, these cities may be marketed and the essences used as a form of inter-urban competition, and civic institutions such as football clubs are transformed by European Union regulation. As such, there is now a substantial literature that seeks to document the dispossession – cultural, political, psychological – felt by long-term residents of cities experiencing radical urban change and population shifts.

So, European cities are symbolically very powerful. However, this focus on cities with strong civic identities neglects the fact that these 'walled' cities no longer compare with metropolitan reality. The city – always divided socially – is now increasingly defined by the growth on its edges, its peripheralised social groups, its non-places and its speed geographies of airspaces, motorways, and high-speed trains discussed in Chapters 5 and 7. The over-riding message of this chapter is that cities are essentially conflictual places, yet retain a symbolic strength that remains vital to debates on the nature of Europe.

5
Travelling Europeans

The first few chapters of this book have been concerned with fixity, and the way in which place identities are constructed. The emphasis now shifts towards mobility and movement, and the idea that Europe can be understood as being a series of spaces of encounter, and a network of technologies. This chapter has four sections. First, some of these mobile practices are virtual – in the sense of being image-based or cyberspatial. Many involve the active consumption of an 'idea' of Europe through a shared language of communication that allows conferences or travel to proceed smoothly. Here, then, the role of the English language in a multilingual continent becomes of fundamental importance. Second, I discuss the importance of corporeal mobility. By this, I mean that while we can point to a map and identify Europe, or we can complain about it or celebrate it as an institution, in many ways Europe is brought into being, brought alive, by a series of practices of mobility, of which tourism is a particularly powerful example. The mobility of these groups is, however, impossible without a whole range of inhuman objects and technologies, and in the final part of the chapter I refer particularly to the development of new mass transport networks in the EU in recent years, and consider how a 'European consciousness' may be born of mobility.

Discursive mobilities

> Why a language becomes a global language has little to do with the number of people who speak it. It is much more to do with who those speakers are. Latin became an international language throughout the Roman Empire, but this was not because the Romans were more numerous than the peoples they subjugated. They were simply more powerful.
>
> (Crystal 1997: 5)

I noted in Chapter 1, one of the most formidable barriers to forging a European identity was the absence of a common language. However, just as the elites of medieval Europe communicated through Latin, a universal language of science, religion, and education, so English is becoming similarly dominant in the New Europe. Part of this can be traced back to the Americanisation of the nation-state discussed earlier in the book, part to the more generalised processes of globalisation in trade, communications, and politics. Indeed, at this point it is worth asking whether English is not a *motor* of globalisation, rather than being simply the passive tool that accompanies these other technological forces. Here, we consider the spread of English as a *lingua franca* (universal language). The cultural commentator George Steiner noted this in a conversation with Richard Kearney:

> Kearney: If we could return to the notion of the European mind. You mentioned earlier that the Dark Ages of Europe was a misnomer, and you seemed to imply that pre-Enlightenment Europe was a time when people had a single culture, and that with the *lingua franca* of Latin, they could move across border and boundaries, and enter into some sort of social and political unity?
>
> Steiner: [. . .]. And what is the *lingua franca* now? Anglo-American or American Creole or commercial American which organizes the computers from Vladivostok to Madrid, the language every young scientist has to publish in, and has to know. I see a terrific contradiction, almost a trap. Can there be this new Europe when it speaks American?
>
> (Kearney 1995: 909–91)

So, given its prevalence, can we see English as a 'killer' of traditional languages (Graddol 1996), sweeping away local languages just as corporate capitalism has apparently swept away traditions, ways of life, and cultures elsewhere in Europe and the world? Yet this is perhaps less the case in Europe than elsewhere in the world. There are four reasons for this.

First, in a number of ways, the popular usage of languages such as French and Italian has undergone significant modification. Cultural influences from, particularly, the United States (and Hollywood) have required the adoption of 'foreign' words, which then make up a 'creolised' language such as 'Franglais' or 'Spanglish'. John Ardagh argues:

> As the French economy has opened to the world, so there has been a permanent incursion of English terms into the vocabulary of business and technology, simply because there are no French equivalents, or they are less neat and snappy. Words like *le marketing, le cash-flow, le software,* even *le design* and *le fast-food,* have become common currency in business or daily life ... But the process, admittedly, may have gone too fast and too far, and the French sometimes borrow falsely: English eyebrows may be raised at such quaint terms as *un tennisman, le footing* (jogging) ... Equally ludicrous are some official efforts to squeeze out franglais by fabricating French equivalents ... *le cash-flow* does not convince when turned into *la marge brute d'auto financement,* nor *e-mail* when it becomes *le courier électronique* ... Attempts to gallicize English spellings, as with *le beuledozère,* have fallen equally flat.
>
> (Ardagh 2000: 704–5)

This raises the ire of linguistic purists, seen in the passing of the 1984 Toubon Act by the French government, which banned the use of certain creolised words in public broadcasting and advertising, a decision subsequently overturned by the French Constitutional Court and subsequent governments (Ardagh 2000: 703–5). More positively, franglais can be seen as a form of linguistic creativity and expressiveness, where the desire to communicate subsumes any essentialised notion of a pure language. Second, Wright (1999: 91) draws attention to forms of communication in cross-border regions where a group will speak in one language, but develop their comprehension skills in understanding the other. Italian and Spanish people can interact in this way relatively easily. Third, there are still international contexts where languages such as French and German are the dominant linguistic form. Fourth, to return to the point made above, the existence of a global 'Latin' type language may allow more space for regional languages or dialects to re-emerge, particularly where they have been displaced by the educational and cultural policies of nation-states eager to cement national unity.

What, then, are the areas that English has come to dominate? Here I identify five.

- *Mediascapes:* With the globalisation of news programming, advertising, newspapers and magazines, television and satellite broadcasting, popular music and cinema, the need for a global language of communication becomes an economic issue. Huge economies of scale can be achieved by media corporations such as CNN that can pick up on the advertising revenues of other global corporations, such as Nike or Microsoft. Appadurai (1990) calls these flows 'mediascapes' by which he means both 'the distribution of the electronic capabilities to produce and disseminate information ... [and] the images of the world created by these media' (pp. 298–9). However, most significantly for our purposes here, most 'Europeans' only have a sense of other 'Europeans' through these modes, but even more significantly there is likely to be a significant mismatch or disjuncture between the idea of a coherent European audience and, instead, one based on transnational community and diaspora. Here, Morley (2000) – considering the case of Turks in Germany – notes that 'new technologies such as satellite not only disrupt national boundaries as containers of cultural experience. They also help to constitute new, transnational spaces of experience ...' (p. 168). Appadurai (1996: 4) summarises this memorably: 'As Turkish guest workers in Germany watch Turkish films in their German flats, as Koreans in Philadelphia watch the ... Olympics in Seoul through satellite ... and as Pakistani cabdrivers in Chicago listen to cassettes of sermons recorded in mosques in ... Iran' so we have a whole new public space, one that is defined by transnational links rather than the sharing of territorial 'containerised' space.

- *Business, diplomacy, and professional communication:* As Goodman (1996) puts it, 'market forces speak English'. While this has produced a whole new lexicon and way of speaking English itself, it also means that in the basic form of business communication, from informal meetings to reports to trade magazines, English dominates (notwithstanding the post-colonial links between, say, Spain and the countries of Latin America, or France and its former colonies). Advances in information technology means that the headquarters of a transnational corporation such as Ford can communicate information instantly to its vast network of employees spread around the world. With the rise of supra-national institutions, such as the United Nations, NATO, or the EU, the diversity of linguistic backgrounds means that English is very often chosen as the lingua franca (Crystal 1997: 78–82, provides a fascinating summary of the range of bodies in science or in trade – from the European Aluminium Association to the European Academy of Anaesthesiology to the European Association of Fish Pathology – that only work in English). Furthermore,

Crystal notes that, in the EU particularly, the problem of of finding translators that can pair languages such as Greek to Finnish means that English is used as an *interlingua,* a language bridge, in other words.

- *International tourism and travel:* The need to communicate safety instructions to an international audience is clear, and as Crystal (1997: 96–8) notes, most transport safety signage is in English. From shipping routes to air traffic control, unambiguous and standard forms of English are used to avoid potentially fatal miscommunication. However, despite the ability to master 'Airspeak' – the limited code of communication with the ground – many pilots would struggle in other English language environments. Furthermore, the idea of standardised non-places such as airports, tourist zones, or major hotels are given uniformity by a number of factors, such as air-conditioning, design, but also – perhaps above all – by the predominance of English in signs, menus, or tannoy announcements.

- *Education:* English as a Foreign Language teaching has mushroomed in recent years. Furthermore, particularly at higher level university communication, such as in the publication of research papers or international conferences, English is unequivocally the language of choice.

- *Cyberspace:* Yates (1996) identifies three principal issues relating to the use of English in cyberspace. First, the Internet's dominant language was based on the American Standard Committee for Information Interchange (ASCII) for many years. This meant that electronic transmission of documents between language communities that use, say, characters with accents (which applies to all the main European languages excluding English) has been very problematic. Second, communication across national groups on, say, mailing lists and chat groups is likely to be in English, again due to problems in communicating in non-Romanised alphabets such as Japanese. Third, there is the transformation of the English language itself through the increased speed at which texts are written and communicated, leading to greater informalisation, abbreviation, and the development of hypertext links. While this is of primary interest to literary and linguistic theorists, it is nonetheless of great importance for the future shape of the language.

Corporeal travel

> People sometimes wonder whether travel and the ensuing encounter with other cultures is, indeed, a factor of cultural integration. It is true that someone who travels does move to locations where people live and behave according to different norms, creating a different culture. But do not most travellers carry only themselves, namely their own identity and their own prejudices which are, consequently and sometimes even agreeably, confirmed by confrontation with the other? Often, travel does not seem to lead to positive interaction, let alone integration.
>
> (Rietbergen 1998: 259)

Central to the idea of a Europe of flows is the movement of *people.* Even in the European context, this is hardly a new concept. The continent is characterised – even defined – by centuries of marching armies and various migrations, forced or voluntary. What is perhaps new is the growing ability of many national citizens to be able to afford temporary visits to other countries, and a growing access to information and cultural knowledge about other countries. Here, tourists and students take their place among the growing groups of business, professional, and economic 'transnationals' who may permanently or temporarily position themselves in other national territories. We might try to categorise them in the following way, bearing in mind one key question: are these migrants and movers undermining a 'European' identity, or are they promoting a European consciousness?

Economic migrants

In the decades following the Second World War, the reconstruction of Europe was based upon an expansion of manufacturing that exhausted the labour supplies of the West European countries. This shortfall was made good by a variety of schemes which brought – usually male – migrants from the Mediterranean region to Western and Northern Europe. This pattern varied by country, but West Germany was notable for its guest-worker (*gastarbeiter*) scheme, where agreements were made with second countries – particularly Turkey – to give work permits to temporary migrants. France, the Netherlands, Belgium, and the UK built upon ex-colonial ties which allowed varying degrees of permanency to migrants from these countries.

However, the broad-based decline in manufacturing, and a more general end to the three decades of economic growth which had accompanied post-war reconstruction, meant that labour markets were no longer stretched, and governments increasingly began to introduce restrictions on migrants. Yet within the global economy as a whole, a greater turbulence – even chaos – has meant that the pattern and flow of migration has altered. As King (1997: 22–3) suggests, the new migration patterns that have come about through an intensified capitalist globalisation are fivefold: first, the globalisation of the world economy means that migrations come from an ever more *diverse* range of cultural and ethnic backgrounds. No longer are migrations simply from ex-colonies; second, as a result, migration is more *differentiated* in terms of social class, with a 'greater variety of *types* of migration', from those on long-term contracts, to refugees, illegal migrants, highly skilled professionals, and commuting migrants who may stay a very short time; third, there is an *acceleration* of migration, in terms of sheer numbers of people crossing borders; fourth, there is a growing *feminisation* of migration. By contrast with the waves of the 1950, 1960s, and 1970s, what is happening more recently is an increasing number of women moving to, say, Spain and Italy in search of domestic work. Finally, the classic 'push' and 'pull' explanations of migration have changed. The end of the shortage of manufacturing labour that was a major pull during the Fordist boom years means that the new service sectors increasingly stratify workers, with highly skilled professionals being well paid, but with an increasing casualisation of work at the lower end of the scale, in tourism and agriculture, for example. Similarly, it could be said that the growing environmental problems in developing countries, along with political unrest and greater knowledge of the North and West are increasing the 'push' towards the richer parts of Europe.

However, while the growing demands of a globally excluded 'working class' often – with good reason – dominates discussion of migration flows, there are also other flows of migrant labour which have to be theorised in the context of European integration.

A professional and business class

In an increasingly integrated Europe, there has been a growing intensity in European business and professional flows. As noted this is linked to the use of English as a 'global language', where Swedes will communicate with Italians and British in the modern-day equivalent of Latin. Ulf Hannerz (1996: 237–51) describes this group as possessing a cosmopolitanism where multilingualism, cultural capital, and shared transnational communities of interest (such as in 'universal' fields of science, architecture, and art) transcend national boundaries. These will not necessarily coalesce exclusively at a European level either, but nonetheless territorial closeness may facilitate communication and face-to-face contact.

We might also consider the daily travel patterns of businesspeople at various levels of the corporate world, from salespeople, financial auditors, directors of transnational companies, and so on. As Davidson (1998) describes, these movements account for several economically significant activities, from those who travel as a routine part of their job (commercial sales people) to those in transnational corporations who have regular meetings with their counterparts in other countries, to the exhibitions and trade fairs that showcase products or services, in a more explicit way than straightforward advertising or marketing literature allows. Such people might be expected to be among the most ardent or conscious Europeans, given the primacy of economic integration within the European project (Mann 1998: 197).

Celebrities (image-based personalities)

Through the pervasiveness of a globalised media, we are exposed to a culture of travelling celebrities. Many are in some way 'European', such as recognisable politicians (Jacques Chirac), religious figures (Pope John Paul II), royalty (King Juan Carlos II, the Queen), or are drawn from sport (Michael Schumacher, Boris Becker, Thierry Henry), film (Gerard Depardieu), or music (Bjork). Of course, each of these listed celebrities is usually associated with a European nation, or else has achieved global status.

Students and universities

One of the most innovative manifestations of European integration is the growing number of European exchange students following programmes in universities in other European countries. Here, the ERASMUS or SOCRATES initiatives were designed to create a common European experience among its youth, and stressed the potential integrative of a Europeanisation of education. Yet this is not a new phenomenon. Rietbergen notes that in pre-modern Europe, intellectual elites were engaged in complex patterns of mobility, with a cosmopolitan consciousness similar to that of today:

> Depending on their place of origin and their financial possibilities, students for whom a good education was a career necessity or a sociocultural requirement, travelled to the nearest or to the most famous academies. Swedes went to Heidelberg, Leiden or Utrecht if they were not happy with Åbo, Lund, or Uppsala. The Dutch, depending on their religious background, travelled either to the catholic universities of Louvain or Orléans, or to protestant Heidelberg, Geneva or Saumur. The French travelled to Pisa or Bologna. Young Greeks from the Orthodox East flocked to the famous academy at Padua, the university of Venice, before taking up positions in the formerly Christian parts of the Ottoman Empire, or even in Ottoman imperial bureaucracy.
>
> (Rietbergen 1998: 271)

The situation is similar today, but more widespread. As Murphy-Lejeune (2002) describes, students are a specific type of migrant: 'highly skilled workers, seeking professional added value or moving for study reasons, and whose migration may only be temporary' (p. 2). Here, students overlap with all of the other categories considered here, from terrorist to tourist, and as Murphy-Lejeune's ethnography describes, the experience of student travel is likely to be formative in any number of ways, from culture shock, to the creation of a European identity and consciousness, the development of linguistic skills (that are not merely limited to learning English), and the creation of a specific social scene which may be one of the purest examples of a consciously European identity.

Illegal and criminal movements

> Ahead of any other multinational industry, organised crime calculated the enormous potential of the European Union through all its formative stages: the mobs of the world – not just European – were the first to go into business, the very first, in fact, to become true Europeans, even if a lot don't qualify from either birth or nationality.
>
> (Freemantle 1995: xiii)

A fundamental drawback of encouraging mobility in the EU is the fact that some of the most determined and successfully mobile agents are operating beyond the law. These include terrorists, drug dealers and couriers, spies, and those travelling without papers. Each of these groups have a substantial material presence *and* a significant grip on the popular imaginary, whether in the cold war novels of John Le Carré, or in the fears of silent invasions of asylum-seekers in various countries. Manuel Castells (1997b) – sees this as part of a more general phenomenon of the 'powerless state', where democratic institutions might be undermined by organised crime. This includes the penetration of the state – either at a local or national level – through corruption, unregulated or illegal political funding; the growing significance of global crime management in international relations; or the impact of criminally-driven financial flows on national economies. Most notable here would be the intense networks of mafia influence in Italy that led to the collapse of the political system in 1993.

Freemantle's (1995) grimly witty assertion of Europe's criminal citizenship draws attention to an important geographical issue: the porosity of Europe's links with the 'outside world', and the increasing impact of criminality on states across Europe by nationals of non-member states. So what are the main areas of crime that are affecting contemporary Europe? Freemantle's (1995) account is somewhat terrifying, including illegal arms trading, drug trafficking, prostitution rings, and white collar crimes such as art theft and computer fraud. Furthermore, as Europe's 'fortress' borders are raised ever-higher, so illegal people-trafficking proliferates, where criminal operators smuggle in desperate migrants through over-loaded boats or, notoriously, in refrigeration lorries or as airplane stowaways. These activities are further tied to money laundering operations by which criminal organisations can remove any trace of their involvement in the illegal transactions: drug trafficking is only profitable if the cash proceeds can be laundered successfully. For example, Driessen (1998) provides an ethnography of the perilous nocturnal boat trips of paperless North Africans across the Straits of Gibraltar, followed by an overland journey to the major cities where, in the likes of Madrid's Plaza del Sol, they might be reunited with friends or contacts. This illegal entry is often facilitated by organised criminal gangs who charge the would-be migrant astronomical sums for their passage.

Tourism

Tourism as a practice in the diffusion of cultural flows has particular salience in the European context, because aside from the huge numbers of Europeans who travel around the globe, many of these journeys are made from, say, Germany to Spain, or from Britain to Italy. As such, they provide most Europeans with their only first-hand knowledge of their apparent co-citizens in any future European polity. Through the fairly mundane process of tourism, then, we might glean some clues as to how such a culturally diverse set of peoples might interact. As Urry (1995) has suggested, tourism is one of the key forms of international interaction on a daily level, for two reasons. First, it encourages *familiarisation* with other countries and cultures, albeit in a deeply artificial and problematic context. Second, it may generate a *cosmopolitanism* among some travellers, an echo of the grand tours of the British upper classes at the beginning of the nineteenth century. In both senses, however, there is a clear degree to which the European consciousness so craved by integrationists may indeed be cradled through the practices of travel and tourism.

European tourism has a long history, usually elite-driven, now within the reach of a mass market. Spas, ski holidays, pilgrimages, the 'grand tour' of the nineteenth century, seaside holidays, second home ownership — these are by no means a new phenomenon.

Yet despite increasing scholarly interest, there is immense scope for more research on the actual nature of this, because if there is any possibility of European cultural under-standing it surely emerges from the myriad journeys made by football fans, church groups, civic associations, au pairs, Erasmus and Leonardo student exchanges, inter-rails, or the modern-day equivalent of what Sampson (1968: 243) calls the 'neo-tourism' of hitchhikers and buskers, or else the glamour circuits of the Cannes and Venice film festivals, yachting marinas, and high class hotels.

Of course, tourism is one of the principal forms of corporeal mobility existing in Europe. This is of a variety of types.

- First, as noted in Chapter 4, there is the idea of a 'cultural tourism' to cities such as Florence, Salzburg, or Paris, that in many ways cements the idea of a European 'heritage', the association of a range of cultural artefacts and producers — Michelangelo, Mozart, Rodin, for example — with a 'European' high art canon.

- Second, by contrast, the growing importance of Barcelona, Antwerp, or Glasgow as tourist destinations is evidence of both the appeal of an 'alternative' tourism, driven by an avant-garde or contemporary — rather than fossilised — art scene, one that is extended to areas of fashion, modern architecture, or nightlife.

- Third, there has long been a countervailing desire to escape from the bustle of hectic city life. Tourist boards and travel writers are keenly aware of this desire, where 'traditional' ways of life can be packaged into a literal fiction, a thematised version of the local culture, be it in Provence or Ireland.

- Fourth, to emphasise Fraser's (1998) complaints, this can encourage a lazy, apolitical view of the on-going processes of underdevelopment or, alternatively, commodi-fication of the history of these areas. Many of these processes are embodied in the 'theme park', or heritage industry, both as metaphor and as real space, where European history is twisted, sanitised, and represented in a programmed, secure environment. The argument that Euro Disney is the future of European tourism is unlikely, but it does provide an excellent example of the disembedding nature of the tourist practice — history lessons without too much attention to detail (Halewood and Hannam 2001 provide a more upbeat account of Viking heritage tourism).

- Fifth, there is the problem of visiting sites of huge loss of human life, a 'dark tourism' (Lennon and Foley 2000) which treads a narrow line between respectful remembrance and a search for understanding on the one hand, and a voyeuristic consumption of death and disaster on the other (a dilemma faced by tourists to the ex-Yugoslavia).

One of the problems faced by EU strategists that was discussed in the opening chapter was the issue of creating a European consciousness or identity. Unlike nation states, which tend to have a reasonable purchase in popular geographical knowledge, Europe is an abstraction, a basket of images, prejudices, and stereotypes. It is heavily mediated, furthermore. And as tourism as a practice is also heavily mediated, it would appear that it would be a prime area for a popular geography of other places to be constructed. Furthermore, tourism is inherently spatial, and tourist spaces are *organised* by a number of different cultural professionals. Even before one leaves home, domestic travel and transport corporations are acting. First, the completion of the single European market has encouraged a change in corporate organisation, with tour companies, travel agents, and tourism becoming increasingly competitive and professionalised. Second, along with the increasing mobility brought about by mass car ownership, European air travel has been transformed in recent years with the growth of charter and budget flights. The rapid expansion of a trans-European high speed rail network, including a direct link to Britain through the Channel Tunnel, has built upon the popularity of inter-rail tickets and other packages (Davidson 1998; Urry 1995: 168–9).

However, these patterns are also affecting place identity and the actual landscapes and built environments of European places. The resort developments aimed at mass package tourism of the Spanish and Italian coasts in the 1960s are perhaps being replaced by a more sophisticated set of holiday demands, though have been replicated in Turkey and East European resorts such as the Black Sea. There is an increasing investment in preservation of historic sites, conservation of significant buildings, and an interest in higher end tourism markets. Furthermore, there has been a clear diversification of tourism types, such as renewed city tourism in cities as unlikely as Glasgow and Naples, an expansion of backpacker infrastructure, business tourism, short-breaks, EU support for peripheral regions to develop 'sustainable' tourist strategies, and so on (Davidson 1998; Urry 1995: 168–9).

As Hughes (1998) suggests, tourist spaces are *produced professionally* (imagined, if you like), through a number of material, visual media. Three are particularly important here in the construction of popular geographic knowledge. First, there is the guide-book, from Baedeker to Fodor to the more contemporary examples of the *Rough Guide, Lonely Planet,* or *Time Out.* Second, there is the advert, where in newspapers, magazines, and billboards, all the highly-developed techniques of the advertising industry are brought to bear in selling places, rather than commodities. Third, there is the tourist brochure which eviscerates place by concentrating on, for example, myths of instant gratification such as food, sea, and sun (see Markwick 2001). Each of these representational media are to the fore in the main forms of tourism in Europe: urban tourism (discussed in the previous chapter), the mass package holiday, and the rural 'escape'.

Mass tourism and the culture of the suntan

The development of cheap air travel from the 1960s, combined with the need for largely non-industrialised regions of Spain, Portugal, and Greece to develop economic sectors, has been one of the most significant cultural trends in European society. High sunshine levels, cheap food and drink, abundant beaches, and easily consumable 'traditional' culture have meant that areas such as the Greek Islands, the Algarve, and the southern and eastern Spanish coasts have become huge draws for working classes from, particularly, Britain, Scandinavia, Netherlands, and Germany. Seasonality of employment, ugly and unplanned resort development, road safety, consumer protection have all been downsides to this image. Hooper (1995: 21) notes that between 1959 and 1973 visitors to Spain rose from under 3 million to over 34 million. In certain parts of Europe, particularly the Greek islands or the Balearics, the impact of tourism has been so significant, that local authorities are looking for ways to curb the worst excesses described above.

The motivation of the Northern European travellers such as the 'Brits on the Costa del Sol' described by Karen O'Reilly (2000) is, clearly, a desire to enjoy a higher standard of living. Yet undoubtedly the motivation for many package tourists can be captured in one phenomenon: the suntan. As Rietbergen notes, most nineteenth-century Europeans on the 'Grand Tour' tended:

> to avoid the summer heat of the Mediterranean, travelling in the relatively milder autumn, winter and early spring periods. When in southern France or Italy, they protected their faces, shielding under huge parasols. Not only did they come from societies with a strong agricultural component, wherein a fair complexion traditionally distinguished the rich landlords from the poor peasants and, consequently, had become a sign of beauty both in the female and in the male, but also they wanted to stress their difference from the darker-skinned 'locals', whose ancient culture they admired but whose contemporary lifestyle was considered so much less civilized and progressive than the society of the democratic, industrialized north-west.
>
> (Rietbergen 1998: 438)

By contrast, by the 1960s the possession of a suntan was a marker of relative affluence and became fashionable. Yet many thousands of the British (and other Northern Europeans) that flocked to the southern Spanish coasts extended this sun worship by adopting the expatriate lifestyle and settling permanently. Here, as O'Reilly's fascinating anthropology demonstrates, they were still largely in possession of the 'less civilized and progressive' attitude to their travel destination. But they were no more in favour of their homeland, instead being left to 'symbolically dangle betwixt and between two cultures and two countries' (O'Reilly 2000: 166). Such transnationalism is shared in a very different context in my next case study – the English in Provence.

Rurality and simplicity: Peter Mayle's Provence

Some of those who do not wish to escape to the beach have sought solace in the rural landscapes of Europe. Here, the likes of Tuscany, rural France, and the west of Ireland have emerged as 'territories of escape' for those jaded by metropolitan life in Europe's large cities (see Chapter 2). In 1989, Peter Mayle first published *A Year in Provence,* an account of selling up and dropping out of the London rat-race and buying a rural farmhouse in Provence. His timing was exquisite. With the southern English economy booming, indeed overheating, there were no shortage of takers with both the money and desire to escape for Mayle's fantasy of the good life. By the end of 1992, *A Year in Provence* was still a best-seller, and together with a sequel, *Toujours Provence*, Mayle ratcheted up half a million sales in 1992 alone (Aldridge 1995), that symbolic year of European integration. Television serialisation followed on the BBC, and the Sainsbury's supermarket chain launched an eponymous *Côtes de Lubéron* wine, with Mayle's signature on the label.

What was the secret formula? In many ways, Mayle was merely feeding an on-going process of second-home purchase by Britons in France. As Buller and Hoggart (1995) argue, the degree to which the French countryside has been affected by industrialisation is far less than in Britain, and affordable dwellings were readily available. Furthermore, one could argue that he offered a competing set of cultural values to those present in an increasingly commodified southern English culture:

> Mayle describes a lifestyle that is simpler and less developed and complicated than that which he (and his assumed audience) have experienced in England (and other, modern urban societies). His French countryside is run not by agri-businessmen engaged in industrial farming but by peasants whose ties to the land go back for generations, who hate to see waste, who do things by hand, who wear traditional attire, and who have good, honest values.
>
> (Sharp 1999: 206–7)

Crucially, he also holds tourists in low regard, ignorant of the 'timeless' ways of the countryside, endlessly prone to misunderstanding and justified exploitation by the cunning locals. As such, he appeals to a knowing post-tourist, similar to Urry's 'cosmopolitan' alluded to above. Furthermore, he captures some of the dominant consumer tastes of the time. As Aldridge (1995: 417) suggests, Mayle's writings 'are to be seen as lifestyle guides to the English. Early retirement, travel, home-building, food and drink: all are major themes' (and it should be noted that Mayle was an advertising executive prior to his success as a writer).

Yet, this fantasy typically ignored the very real problems of Provence as a peripheral region within France. Mayle, writing a new preface to *A Year in Provence* in 1989, seems partially aware of the tension involved in seeking to flee from modernisation, and the needs of a region that has rarely known affluence:

> The only cloud on our horizon at the moment is a small one ... There are hideous rumours of what our local friends call *un boum*. By 1992, they say, Provence will be firmly established as the California of Europe. I hope they're wrong. It conjures up visions of diet-crazed water drinkers in double-knit pastel jogging ensembles, of cordless telephones by the pool and authentic Provençal-style jacuzzis next to the tennis court. I have a terrible feeling that the French would love it all. It's the refugees who hope that Provence will stay the way they found it.
>
> (Mayle 1990: preface)

This is a recurring problem in any discussion of development around the world. In the context of the New Europe, however, it clarifies the tension between core and periphery that exists both in Provence and in Ireland, the desire of those in the most economically dynamic regions of Europe to also have the right to consume the continent in its pre-modern state.

So when considering the effects of tourism on European place identity, there are usually two, apparently contradictory, disembedding processes said to be going on in parallel – a move to a 'placeless' global village of corporate standardisation in entertainment, food, and social practices (e.g. commuting), whether formed by US, Japanese, or European corporate practices; and an *exaggeration* of place essence, witnessed in the selling of rural romance. However, if one is searching for clues as to who is doing the travelling, it is not entirely clear. Package holidays, hitchhiking, student exchanges, caravanning, football weekending, stag parties are not normally associated with the stereotypical 'high art' tourist alluded to above. It is clear that there are a vast number of diverse tourism practices being carried out, and that these are generally under-researched.

Thus there is a range of corporeal movements taking place across Europe, as part of a broader trend associated with more general processes of globalisation. These movements are diverse, the travelling aims of the actors based on vastly differing desires as to length of stay, motivations, and economic and social profiles. Yet what is clear is that these corporeal movements would be near impossible without the technological means of propulsion that have transformed our understanding of Europe and the world over the past century or so. So it is to the geographies of space and time, of speed and movement, that I now turn.

Geographies of speed and movement

So far I have given a sample of various social groups that are regularly moving through European space, deriving this from Urry's (2000) call for a new sociology of flows. However, in many discussions of globalisation, which are often polarised between those who see the future as optimistic (cosmopolitan 'borderlessness') or pessimistic (neo-medieval), there is an added dimension to the debate – the role of the inhuman object:

> Human powers increasingly derive from the complex *interconnections* of humans with material objects, including signs, machines, technologies, texts, physical environments, plants, and waste products. People possess few powers which are uniquely human, while most can only be realised because of their connections with inhuman components. The following inhuman developments are novel in their ontological depth and transformative powers: the miniaturisation of electronic technologies into which humans are in various ways 'plugged in' and which will inhabit most work and domestic environments; the transformation of biology into genetically coded information; the increasing scale and range of intensely mobile waste products and viruses; the hugely enhanced capacities to simulate nature and culture; changing technologies which facilitate instantaneously rapid

> corporeal mobility; and informational and communicational flows which dramatically
> compress distances of time and space between people, corporations, and states.
>
> (Urry 2000: 14)

In this section I consider how the inhuman devices associated with rapid corporeal mobility – cars and motorways, rail networks and airspaces – might open up some perspectives on thinking of Europe as being *constituted* by non-human *things*. I preface this, however, with a discussion of how the European Union contains within it a basic notion of mobility.

The European Union and mobility

> Transport is crucial for our economic competitiveness and commercial, economic and
> cultural exchanges. This sector of the economy accounts for 1000 billion euro, or over 10
> per cent of the EU's gross domestic product, and employs 10 million people. Transport
> also helps to bring Europe's citizens closer together, and the Common Transport Policy is
> one of the cornerstones of the building of Europe.
>
> (Loyola de Palacio, European Commissioner for Transport, in Commission for the
European Communities 2001: 2).

The European Union's motor, its mission statement, has always been premised upon a model of capitalist development that stresses the inefficiencies of nationally-bounded markets. As such, its stress on mobility involves reducing barriers to the free movement of people, goods, and – by extension – the actual things that move them, be they planes, trains, or articulated lorries. To achieve this, the European Commission has sought to develop strategic spatial planning initiatives that help to ease transport bottlenecks and provide new, quick links within the post-national 'macro' regions of Europe – the Atlantic Arc, for example. It has produced numerous policy statements and funding mechanisms that have pursued this aim – the 2001 European Spatial Development Perspective (ESDP) and the *Europe 2000 (+)* reports that preceded it, or the recent Transport White Paper, for example (Commission for the European Communities 1991, 1994, 1999, 2001, 2002). As I discussed in Chapter 1, this involves a reading of Europe where national borders are seen as protectionist barriers, which must be superseded.

More specifically, the EU has prioritised a number of projects to further this goal. In its *Trans-European Transport Network: Ten-T Priority Projects* (Commission for the European Communities 2002), a whole range of projects are highlighted, as figure 18 shows.

These include air, rail, road and, to a lesser extent, water-based transport, and range from the expansion of Malpensa airport (outside Milan), to the improvement of Greek motorways, to dedicated freight lines, such as that linking the port of Rotterdam to the existing German–Dutch rail network. New high-speed train lines, such as the Madrid–Barcelona–Perpignan route, or the North–South Berlin–Verona link, aim to slice through long-standing national networks. Intriguingly, funding has also gone to the new Denmark–Sweden fixed bridge link, creating a striking new cross-border region.

Priority projects adopted in 1996

1. High-speed train/combined transport north–south
2. High-speed train PBKAL (Paris–Brussels–Cologne–Amsterdam–London)
3. High-speed train south
4. High-speed train east
5. Conventional rail/combined transport Betuwe line
6. High-speed train/combined transport; France–Italy
7. Greek motorways; Pathe and Via Egnatia
8. Multimodal link Portugal–Spain–Central Europe
9. Conventional rail link Cork–Dublin–Belfast–Larne–Stranraer (completed)
10. Malpensa airport, Milan (completed)
11. Øresund fixed rail/road link between Denmark and Sweden (completed)
12. Nordic triangle rail/road
13. Ireland/United Kingdom/Benelux road link
14. West coast main line (rail)

Priority projects proposed by the European Commission in 2001 (new projects and extensions)

New projects

15. Global navigation and positioning satellite system Galileo
16. High-capacity rail link across the Pyrenees
17. Eastern European combined transport/high-speed train
18. Danube river improvement between Vilshofen and Straubing
19. High-speed rail interoperability on the Iberian peninsula
20. Fehman Belt fixed link between Germany and Denmark

Extensions

1. High-speed train/combined transport north–south (Verona–Naples and Bologna–Milan)
3. High-speed train South (Montpellier–Nîmes)

— Adopted Rail Project
---- Proposed Rail Project
— Adopted Road Project
---- Proposed Road Project
—— Rail Network (in 2010)
—— Road Network (in 2010)

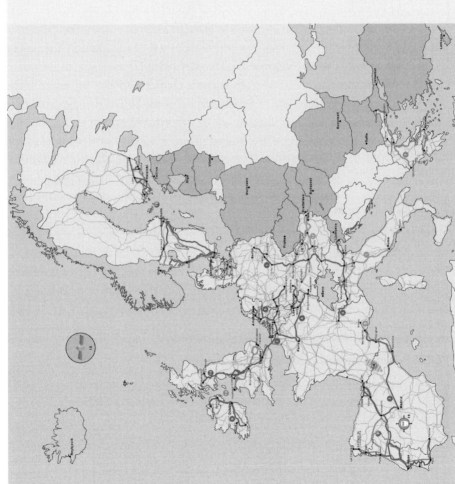

Figure 18 Map of TEN-T projects. © European Communities 1995–2002.

Furthermore, the stress placed on the funding of Galileo – a global navigation and positioning satellite system – is anticipated to improve the monitoring of the movement of people, goods, and vehicles across the continent.

However, while often presented as being a neutral public good, transport and mobility is of course highly politicised. As Richardson and Jensen (2000) argue, there remains an unresolved tension at the heart of the EU 'mobility' documents between co-operation and competition. While the ESDP:

> repeats concerns about 'pump' effects (where new high speed infrastructure removes resources from structurally weaker and peripheral regions) and 'tunnel' effects (where such areas are crossed without being connected) ... all of the policy options pursue the general aim of improving accessibility as a generic response.
>
> (Richardson and Jensen 2000: 513)

Thus the inter-urban competition identified by Harvey (1989) as entailing the creation of a favourable investment environment for corporations will be very directly affected by the new high-speed network (Newman and Thornley 1996). Furthermore, the increased capacity brought about by these links are in direct conflict with calls to reduce traffic both for reasons of congestion and of global warming, as has occurred in the case of the Danish–Swedish bridge link. This situation will be exacerbated once the new member states of East and Central Europe are integrated into European territory (Jensen and Richardson 2000: 514). And as I discuss in the following chapter, the opening and planning of the Channel Tunnel posed huge questions of sovereignty, cultural identity, and economic impact on the regions of Kent and Nord-Pas de Calais (Darian-Smith 1999; Sparke 2000).

So, the European Union's territorial impact will be substantial. Yet the maintenance of such links, and the creation of new networks, will be fundamental to the future cultural identity of a continent that, in the portentous (if exaggerated) words of Loyola de Palacio, has for 'the first time since the Roman era ... started to think about transport systems going beyond national frontiers' (in Commission for the European Communities 2002: 3).

Railscapes and a trans-European consciousness

> Between the dissociating experience of plane-travel, and the frustrations of driving on the roads of Europe, the railway may provide not only a *via media* but a more adaptable and meaningful interpretation of a unified land mass.
>
> (Sampson 1968: 248)

The Orient Express, the Flying Scotsman, the TGV, Eurostar, the Pendolino. The romance of Europe – its diversity, its changing languages, its mountains and coasts and fields – and the sense of arrival offered at Antwerp, at Rome Termini, at London St Pancras, all are part of Europe's powerful railway mythology, and perhaps its railscapes offer the perfect optic for understanding a Europe of flows. The whole idea of the trans-European railway, which has existed since the mid-nineteenth century, demonstrates the functional linkages between the European nation-states, the resolution of linguistic misunderstandings, the provision of shared rolling stock, the problems of international timetabling ... and the sheer exhilaration of watching a continent roll past – the changing signs on buildings, the diverse landscapes – without shifting seats. Yet, as Bishop argues, railways 'pose a theoretical dilemma':

> Do they alienate and destroy a sense of place...or enable new connections and perspectives to be made? If the latter is the case, do these new experiences of place sufficiently compensate the loss of more traditional sensibilities? While many contemporary theorists try to negotiate a path between extremes, it must be said that a sense of loss is pervasive.
>
> (Bishop 2002: 298)

Bishop is discussing the relatively recent construction of the Alice Springs to Darwin rail line across the Australian outback, over 100 years after the completion of Europe's rail network, yet the issues he raises apply to our understanding of the role of rail in the New Europe. Here I want to consider whether, in fact, a European territory is being forged that is not planar, but rather built on speed and time.

By the year 2010, the European Union aims to complete its network of high-speed trains. Here, after a couple of decades of decline, it appears that the railway will return to its prominence as one of the most visible and material expressions of a European territory. While it is still possible, as Morris (1998) remarks, that 'a train from the Peleponnese will eventually unload you in Lapland' or that it is 'theoretically possible to travel on a single ticket from Lisbon to Tallinn' (both p. 244), the new network shuns romance for functionality and speed. The high-speed trains, most successfully pioneered in France, travel at up to 280 kph (Ardagh 2000: 141), and on existing routes have slashed travel times dramatically. Paris to Brussels, for example, which until recently took 2 hours

Figure 19 *Eurostar is part of a new time-geography of Europe based on high-speed trains. Photograph courtesy of Eurostar. Used with permission.*

50 minutes, can now be traversed on the new Thalys trains in 1 hour 20 minutes (Davidson 1998: 131). Furthermore, the advantage of high-speed train over air travel has been recognised, with estimates that journeys under 800 km and under three hours will usually be quicker by train, given the lack of need for check-in or travel to the airports. As high-speed trains can arrive into the heart of old cities, this form of transport has a clear *centrality*. And planners are increasingly working to develop integrated air-rail hubs, where new or existing airports are combined with high-speed rail stations (Davidson 1998: 133).

What, then, does this mean for the idea of a European network? Bearing in mind the comments made by Bishop and Sampson above, does the new high-speed train avoid the 'dissociation' of air travel, or does the speed at which it travels continue to fragment the European territory?

Box 5   The European railway station

The European railway station has undergone a renaissance since the early 1990s. The station essentially functions as a gateway to a city, and as such was subject to the same architectural criteria as a major public building for much of the period from the 1850s to the 1950s (Schivelbusch 1987: ch. 11). With the arrival of the car, their role was down-played, but recent years have seen a reinsertion of their significance within the city. Yet while the speed 'corridor' may imply the idea of non-place, or at least the reduction of spatial difference, new stations are being designed to *enhance* place distinction. And this is undeniably driven by a political discourse of a Europe of networked cities, which plays with the idea of gateways and celebrates the centrality and density that cities offer. Often described as 'cathedrals', we are seeing the refurbishment of existing buildings, or construction of brand new station termini (Parissien 2001).

Governments often employ recognisable 'signature architects' to provide identity and a 'sense of arrival' to these spaces of transience, something that characterised the stations of the golden era of rail travel a century before. Examples include Rem Koolhaas at Euralille, Santiago Calatrava at Lisbon (see Figure 20) and Zurich, Jean Nouvel at Lyon, Nicholas Grimshaw at London Waterloo, and so on. These stations may be placed in previously marginal districts of cities, or they may redefine the city centre. Furthermore, they may act as a new focus or meeting point for the city's diverse publics: 'The collision of different spatial scales (in the most extreme cases from the global of high-speed train destinations to the local of the neighbourhood stations) is mirrored by the presence of a broad range of users (from the cosmopolitan business person to the homeless drifter)' (Bertolini 2000: 461). This parallels the point made by Sudjic (1992), who suggests that airports are the 'new city square', given the mixing of diverse social and ethnic groups in these spaces.

Figure 20 *Calatrava's railway station in Lisbon.* © *Audiovisual Library European Commission.*

Perhaps the clearest indication of what this means is shown in the development of the Euralille railway station. Here, in one of the densest population concentrations of Europe, the leadership of the old industrial city of Lille employed Dutch architect Rem Koolhaas to masterplan a new station area to house a major high-speed rail junction. The intention, as Bertolini (2000) has described, was to come to terms with the new centrality of the city – the station complex was, effectively, to act as a new or parallel city centre for Lille. This was based on two presumptions. First, there is the apparent paradox of the informational society that just as activities are deconcentrated (through new technologies) so the need for face-to-face communication increases. Here, new concepts in office accommodation and business practice are being allied to mobile work practices. This has meant that property developers have increasingly sought to redevelop station areas both in terms of flexible office space and as consumption sites, demonstrated in a range of examples across Europe: Ajax-Bijlmeer in Amsterdam, Frankfurt airport, and Dortmund station (Bertolini 2000: 463). Simultaneously, however, there is also a trend towards refurbishing stations as 'gateways', as in the period of the pre-automobile age. Here, perhaps, new station developments are the catalysts of both the reborn city centres discussed in Chapter 4, and in new decentralised or 'edge' developments elsewhere in the city.

So, perhaps in looking at the emergence of a European high-speed future we can identify a partially deterritorialised Europe, based as much on speed corridors as on a flat map of the network. Here, however, we might consider whether this 'splintering' map of Europe (cf. Graham and Marvin 2001) is in fact reinforcing a kind of cosmopolitan European identity around certain key nodes (and which exclude other spatial areas that remain unconnected to the net).

How might we theorise this? Virilio (1991) in Bishop (2002) suggests that our increasing use of speed has eclipsed and squeezed the territories that we understand. Here the idea of the 'rail corridor' might be used. Imagine taking the TGV from Paris to Marseilles. You cross the enormous regional diversity of France, but the landscape is but a blur. Here, what Virilio calls 'vectors' are slices of fast-time, moving environments (the fixed interior of the train carriage) that transport us through extensive national and transnational territories. However, 'just as the vectorial qualities of the rail corridor (the immensity of its length in contrast with an extreme narrowness, its evocation of mobility, of an efficient journey, its focus on departure and destination) constantly undermine local-territorial or place-related aspects, so too there are moments of intense resistance to, even subversion of, the dominant vectorial definition' (Bishop 2002: 299). So, arguably, railways compressed the idea of Europe as much as any other technology, and long before grander schemes of European integration came to fruition. Could it be said that with the advent of high-speed networks, they will once again vertebrate a spatial imagining of European territory, while retaining a sense of attachment to territory?

Motorways

> For me one of the excitements of Europe is the immense cavalcade of trucks which perpetually criss-crosses the continent, night and day, growing from a tentative trickle at the end of the 1940s almost to saturation fifty years later . . . I see the trucks as fulfilling historical schedules. The roads they take nowadays are mostly freeways, but often enough they keep to the tracks of ancient trade routes, along the same valleys, over the same passes, between the same centres of commerce and industry, crossing the same rivers high above the passages of prehistoric fords or ferries.
>
> (Morris 1998: 237)

Morris draws attention to the intriguing nature of the contemporary European motorway. Far from being non-places, far from whipping us into the modern day as uprooted citizens of a virtual world, perhaps we are seeing simply an intensification of the speed of connectedness. And, furthermore, the physical topographies of contemporary Europe are far from absent, even if they are cut back and tamed and barely visible from a speeding car or lorry. And yet, in the tighter confines of the city region, it is clear that these 'speedscapes' do in fact reconfigure how we understand the European continent (and its urban sprawls). However, 'whereas rail symbolized collective, societal strength, based primarily on the length and breadth of the national infrastructure, the motor car has become a universally recognized symbol of individual wealth and status. Vehicle ownership replaced network infrastructure as the yardstick.' (Ross 1998: 94–5).

While railways may be quintessentially European, what about motorways? There has been a tendency to ignore these crucial structuring spaces in both academic and popular commentary, an issue I pick up again in Chapter 7. Cars, generally, have become a fundamental part of modern life, central to numerous new social practices, from shopping to the school-run, tourism to commuting. Yet they have been curiously under-examined by many sub-disciplines of social science. As Sheller and Urry point out 'civil societies of the West are societies of "automobility"' (2000: 738). As with railways, motorways are a key aspect of a European network of moving bodies. By contrast, however, cars represent a major problem for all advanced societies. According to Sheller and Urry (ibid.), 'it has been presumed that the movement, noise, smell, visual intrusion, and environmental hazards of the car are largely irrelevant to deciphering the nature of city life'. Here, if Europe's Union is to be justified by a high material standard of living combined with greater individual mobility, the negative environmental impacts caused by the car, van, and articulated lorry will have to be tackled on a continent-wide scale. This will not be easy.

Air travel

The network of air transport has been drastically restructured since 1997, when the single European air market came into force, challenging the primacy of the national 'flag carriers' such as Sabena, Alitalia, or Iberia. So how do we measure such flows? As with the development of the rail network, there are two issues to bear in mind: one is the airport which handles the passengers, which may be competing to be a hub within a network of smaller 'spoke' regional airports, handling transfers of passengers; the other is the actual network of routes that may be followed, and which gives an insight into the movement of people within the European continent. For example, in 1996 almost 660,000 people flew between London and Stockholm, and a similar number flew between Stockholm and Helsinki. 865,000 people flew between Paris and Amsterdam, and 920,000 between Milan and Paris (Davidson 1998: 94). Furthermore, as I discuss at the end of Chapter 7, there is no reason why the airport should not be as iconic a place of cultural imagination as the grand early twentieth-century railway stations that appeared at the centre of cities (Parissien 2001). Barajas, Malpensa, Schiphol, Heathrow, and Roissy-Charles de-Gaulle are now at the forefront of an ever-expanding European air network.

Seaports and maritime cultures

From the Atlantic Ocean to the Black Sea, and from the Baltic to the Mediterranean, the shores of Europe are often characterised by an interpenetration of land and sea that has facilitated and encouraged the flowering of many maritime civilisations, as well as a great variety of political and trading systems. Great cities have grown on major estuaries (Antwerp, London, Bordeaux), around fine natural harbours (Genoa, Istanbul, Marseille), or on defensive island sites (Copenhagen, Malta, Venice). Ancient Greeks and Vikings developed extensive trading systems from their coastal ports. Fifteenth-century Portuguese explored African shores, while the Italians and Spanish reached the Americas. European empires developed from these beginnings, based on maritime trade and transport: this imperial system was largely dependent upon major European cityports (Antwerp, Hamburg, Liverpool, London, Marseille) and their counterparts implanted on distant shores (Boston, Dakar, Sydney, Vancouver).

(Hoyle and Pinder 1992a: 1)

Finally, it is crucial to consider the role of the sea and water-based transport in the structuring of a European society. Ross (1998) suggests three issues of significance here. First, the sea and waterways have often been a crucial marker of territorial identities, particularly in countries such as the UK or Sweden with long coastlines. However, major rivers such as the Rhine or the Oder have also acted as significant trading routes or geopolitical boundaries in history. Second, these spaces pose difficult issues for European governance. For example, the Barcelona conference of 1995 sought to establish a common regime for the governance of the Mediterranean, a space shared with a very diverse set of members. Third, very interestingly for our purposes, the sea has always enhanced Europe's cosmopolitanism, with history revealing flows of peoples in, for example, medieval trading empires, undermining any idea of a 'pure' race or society. Here, port cities such as Barcelona, Marseilles, Hamburg or Naples have always been characterised by a high degree of multiculturalism.

Indeed, the port system is a fundamental component of European identity (Hoyle and Pinder 1992b, Hoyle 1996), and is an area where Europe's essentialisms have been constantly challenged. However, they have changed drastically in nature in recent years, due primarily to advances in technology and containerisation. For example:

Liverpool's population in 1994 was estimated at 474, 000, just 60 percent of the 789,000 in 1951. At its peak, the port employed 25,000 dockworkers. The MDHC [Mersey Docks and Harbour Company] now employs about 500 dockers (and sacked 329 of these in September 1995). Similarly, a very large proportion of the dock traffic is now in containers and bulk, both of which are highly automated and pass through Liverpool without generating many ancillary jobs locally ... The warehouses that used to line both sides of the river have been superseded by a fragmented and mobile space: goods vehicles moving or parked on the United Kingdom's roads at any given time – the road system as a publicly funded warehouse.

(Keiller 2001: 448)

The impact of these changes on cities are remarkable. For Keiller (2001), however, the invisibility of ports matches the invisibility of the British manufacturing economy, a situation mirrored in much of Western, Southern, and Northern Europe, certainly.

Conclusion

This chapter has sought to trace out the idea that Europe is – and has always – been based on *mobility* and encounter. To conclude, I make four points.

First, Europe is constituted by an incredibly complex series of flows – economic, technological, financial, ethnic, image-based – that undermine any attempt to construct it as a simple evolution of political sovereignty. While much of this is due to the fluidity of capital flows – be these financial, commodity, legal or illegal, human or inhuman – nor can the cultural identities that exist on European territory be reduced to economic determinants.

Second, these flows are not new, but there is an increasing complexity to their patterning. Theorists of migration have demonstrated that there was a strong correlation between the post-war evolution of the European economy and flows of, particularly, guest workers. Now, however, the situation is far more complicated, and will become increasingly so once the countries of East and Central Europe are admitted to EU membership.

Third, it should of course be borne in mind that mobility, by its nature, escapes the tight borders of 'Europe' as both territory and term. Islamic terrorists, Mossad secret agents, Russian and Italian mafia, may use European territory as a base and theatre for international power-plays that have very little to do with the nations involved. The images of film stars that dominate cinema screens and others media are likely to owe some debt to Hollywood. The dramatic kidnapping of Nazi war criminal Adolf Eichmann by Israeli agents in 1960 took place in Buenos Aires for crimes committed in Europe. The tourists that flock to European cities and beauty spots are from Japan, Australia, the United States, as well as from other European countries. The media corporations that are driving forward the use of English are transnational. So Europe is actively involved in the global, by virtue of the historically constituted networks of power and culture that individual nation-states have forged with the rest of the world (not least through imperial entanglements).

Finally, this should not be seen as a dual process of human flows on the one hand, and technologically driven movements on the other. The two are intertwined – most migrants arrive at their destination by plane or train, whether they be rich or poor. Linguistic communication, especially in English, is perpetuated by a whole complex of face-to-face conversations, telephone calls, faxes, e-mails, or other forms of interaction. To remind ourselves of Urry's (2000) argument, bodily or corporeal mobility is heavily reliant on non-human objects, and the human-object relationship can, and indeed *must* be extended to any understanding of European integration.

From this brief survey it can be suggested that Europe has always been partly constructed by the interaction of human mobility desires and technological forms of mobility. The 'speed corridors' that provide the backbone of the continent are crucial factors in developing an advanced sense of European identity, and they thus demand a greater degree of study than merely considering the EU's transport policy framework. One area of great importance is the degree to which transport and mobility are, of course, regulated by barriers and borders, as Chapter 6 explores.

6
Borderlands and barriers

At the time of the momentous events of the late 1980s, optimistic noises were being made about the post-national version of Europe, the opening up of borders within the European Union, the pulling down of walls dividing the capitalist and state communist countries so artificially created after 1945. At the same time, the development of the Internet, of satellite broadcasting and e-mail, and vast expansion of transport technology had created optimistic visions of a Europe less internally divided than ever before. Here, the theme of mobility and freedom was an important aspect of the debate emanating from pro-Europeans, an attempt to supersede the tyranny of borders. Previous chapters have addressed issues of mobility and travelling, and the uneasy co-existence of European, nation-state, regional, and city spaces and identities. Many of these identities require some notion of otherness and boundedness for their cohesion, which in turn may require the construction of boundaries and walls, at the same time as others are falling. For example:

> The media-generated moral panic, at the turn of the millennium, about 'illegal immigrants' and asylum-seekers 'invading' the symbolic white cliffs of Dover has been one of the most shameful recent episodes in recent British history. It would seem that there are still many who wish that the Channel Tunnel linking Dover to the French port of Calais ran only in one direction – so that British holiday-makers could have access to 'the Continent' when they wished, without permitting any but the most utilitarian, commercial traffic in the other direction.
>
> (Morley and Robins 2001: 8)

In recent years – in Western Europe at least – the idea of an identifiable, militarily aggressive, European enemy has disappeared with the increasing cultural and economic integration of European societies and economies. In this context, the idea of threat or invasion has been targetted at different sources. One of these might be an influx of 'others', such as asylum-seekers. Here, the idea of a unified territorialised community is disrupted by the networks of belonging that many immigrants may maintain with their homelands. A central notion here is the idea of diaspora, 'a persistent sense of community between people who have left their homeland (usually involuntarily) and who may be scattered all over the world' (Castles and Miller 1998: 201). Countries such as the US and Australia are now home to many European diasporic communities, yet in 'old Europe' there has been some resistance from conservatives to the presence of 'others' within their communities.

In this chapter, I address the issue of borderlands – those areas where legally defined territories run up against each other – and their transformation in the New Europe, and at the same time consider the nature of barriers to mobility that are being erected. This has been one of the fastest growing areas in European studies in recent years, spawning a whole range of works both theoretical and empirical (see, for example, Anderson (1996), Anderson and Bort (1998), Bort (1998a), Kaplan and Häkli (2002), O'Dowd and Wilson (1996), *Space and Polity* (2002), Spohn and Triandafyllidou (2003), Wilson and Donnan (1998a)).

To pursue this, consider the following assertion:

> The frontier is never 'natural' or a matter of physical geography. It is always a political process – an institution defining difference with the outside world and attempting, by influencing mentalities, to homogenize the diverse population inside the frontier. It is therefore a political 'technology' which records the balance of power at a particular time in space.
>
> (Bigo 1998: 149)

This chapter takes issue with – though does not dismiss – this vision of a less divided continent. I consider three themes. First, I set out the idea of 'Schengenland' (Walters 2002a and b) as a new territorial formation that accompanies the development a political idea of the EU. Here, the idea of fixity and flow, and the tensions involved in seeing Europe as being a *mobile* society is challenged. Second, I consider different concepts of the border, as (a) physical, (b) networked, (c) narrated, and (d) political-technological. Third, I consider the emergence of transfrontier or cross-border regions such as the Channel Tunnel and a variety of cross-border projects that are being developed all over the new European territory, and consider the way in which bridges are recurring metaphors in the language of European integration.

Schengenland and Europe's external borders

Throughout this book, the emphasis on European integration as involving freer movement has been recognised. Yet this has come with an increased fear of the illegal movements identified in the previous chapter, and a more general political hostility towards migration from outside the EU. The issue of borders has often been presented as being a neutral thing, often playing upon the significance of natural or physical borders. For example, Foucher (1998: 242–8) usefully summarises the external frontier of the EU in the following way:

- The Maghreb, separated from the EU by the Straits of Gibraltar.

- Italy, which has a land border with Slovenia and an Adriatic coastline which faces Albania.

- Greece, to the south-east, shares borders with Albania, Macedonia, and Bulgaria. However, the biggest issue concerns its relation to Turkey (and relatedly, Cyprus), with which it has a long-running enmity.

- Austria: with land borders with a number of nation states seeking EU membership, the Austrian frontier is a key area in the construction of an idea of Mitteleuropa (see Chapter 1).

- Germany, which has to its east and southeast four states: Poland, the Czech Republic, Slovakia, and Hungary.

- The Baltic Sea to the north, which is developing as a new economic area in the context of the post-1989 political landscape. Here, states such as Estonia are developing new links with their neighbours, Finland particularly, and seeking to Europeanise themselves after a long period of Soviet domination.

- Finally, and similarly, the accession of Finland to the EU means that the Union now has a 1313 km long frontier with Russia, thus opening questions as to the future relations with the still important power.

However, Foucher (1998) is concerned to show that these borders are highly politicised. The issue of borders is central to the European integration project, as a key aspect of the dominant agenda in governance (the economic) has been the *removal of barriers* to trade. On the flipside, there has been a more positive move to encourage mobility, through the creation of technologies which lubricate movement. In particular, the European Commission has driven forward the creation of a European high speed train network and has encouraged the consolidation of a European motorway network. It has created a new funding programme, Interreg, which has given significant financial support to projects that seek to improve porosity between member states, both in tangible cases (such as new bridges), or in attempts to enhance social integration (such as cross-border labour market studies).

Yet as the discussion of the Europeanisation project suggested, the search for internal permeability (itself a controversial notion due to fears of freeing movement for terrorists, smugglers, and poor migrants) is tempered by an increasing premium on policing the borders of the European Union. Thus the strength of the borders on the Spanish–North African coast, Italy's Adriatic shores and the lines of the old iron curtain have become of increasing concern to politicians as a means of restricting the mobility of those who wish to enter Europe. As I discuss below, this has given rise to the metaphor of 'Fortress Europe', the idea that just as the internal community is strengthened, so outsiders are increasingly viewed as hostile invaders, who must be excluded through physical defence.

Fortress Europe and Schengenland

Westphalia, Vienna, Versailles, Potsdam, Maastricht . . . if the history of Europe's formation as, and within, a space of territories, sovereignties, economies and cultures can be evoked in terms of such symbolic place names then perhaps we can add to that the name of Schengen.

(Walters 2002a: 561)

Schengen, in Luxembourg, was the place where in 1985 the nation-states at the heart of the European territory met to sign an agreement concerning the movement of people across their borders. Walters is surely right to place it in the same category as the other major sites of treaty-signing in the last 300-odd years of European political geography, for the agreement reached at Schengen has had, and will continue to have, a massive impact on how Europe is constituted. Furthermore, in any consideration of Europe as a *mobile* society, the creation of 'Schengenland' is an explicit shift away from borders as being at the margins of national territory, towards one where they are *central* to the construction of a European identity. And in adopting this geographical perspective, there is a very clear shift away the orthodoxy of international relations (IR), which tends to assume that the state is an unproblematic starting point for research. In this section, and drawing on a number of commentaries (Newman and Paasi 1998; Walters 2002a, b) I discuss various perspectives by which this has been rethought.

In 1985, members of the EU agreed radical new measures to transform the structure of borders between member states. This – the Schengen Agreement (extended in 1990 and included in the Maastricht Treaty) – was driven primarily by Germany and France, whose internal borders with the Benelux countries were increasingly meaningless. In principle, Schengen rested on two pillars: first, it created the foundations for a common European security policy, directed primarily against organised crime, drug and people trafficking, and illegal migration; second, it eased the basic principle behind the Single European Market that people, as well as goods and services, be supra-national. As in the neo-liberal ideology that underpinned the single market's operation, so it was felt that labour had to flow to areas of demand, and abolishing borders was seen as central to this (Walker and Travis, 1998).

However, the agreement was framed without the foreknowledge of the impending collapse of the communist system, and by the mid-1990s, the turmoil in Eastern and Central Europe, along with the demise of Yugoslavia, sent thousands of asylum seekers over the borders into the EU. Along with other major political migrations, particularly Kurds fleeing persecution from Iraq and Turkey, and Albanians fleeing the collapse of their economy, huge pressures have been placed on the member states' asylum systems. Given the nature of this migration, there is now a very significant issue involving the internal nature of borders, the idea that the physical frontier (based on the notion of blockade at territorial borders) is being replaced by a functional frontier (based around surveillance and monitoring of societies) (Foucher 1998).

Concepts of the border

The physical barrier

Traditionally, a lot of attention has been paid to the question of nature and the physical environment – in particular the historical symbolism of river or sea boundaries, and the impact of ecological change on the significance of such boundaries. For example, the

Mediterranean has frequently been compared with the Rio Grande along the Mexican–US border, but this divide is culturally and economically even sharper than in North America. King (1998: 110) identifies five clear contrasts between the societies on either shore of the Mediterranean:

- As an economic divide, marking the so-called first world from the so-called third world;

- As a demographic divide, between the high-fertility regime of the southern shore contrasting with extremely low birthrates in Italy and Spain;

- As a geopolitical divide 'between stable, democratic Europe and the generally less stable and less democratic regimes of North Africa and the Middle East' (King 1998: 110);

- As a cultural contrast between Christian Europe and the Islamic south;

- And as a migration frontier between areas with a strong 'push factor' and the desirable European territory.

Braudel (1976) has famously argued that the Mediterranean between Spain and North Africa acted in medieval times as a river, rather than a barrier, uniting, rather than dividing, the two land areas (Driessen 1998). From the early 1500s, however, after the expulsion of the Moors from Spain, the Strait of Gibraltar became a political frontier. Today, the Mediterranean thus functions as a frontier in a number of ways. Economically and demographically, there is a clear divide between the economic growth rates and birthrates of the two sides of the 'river'. However, there is also a symbolic divide: 'an ideological and moral frontier, increasingly perceived by Europeans as a barrier between democracy and secularism on the one hand, and totalitarianism and religious fanaticism on the other.' (Driessen 1998: 100). Increasingly, the narrow stretch of water – and the Spanish enclaves of Melilla and Ceuta within North Africa – has become a frontier similar to the border area between Mexico and the US.

The border as networked

So, a key idea of the border is as obstacle, a physical 'fortress' where through natural or built defences, anyone trying to enter sovereign territory who is unwelcome may have to physically pass a border (perhaps using illegal means – think of the 'shadow' mobilities of Chapter 5). Yet increasingly the border has to be seen as something 'deeper', more complex. As Walters (2002) explains, Schengen has removed some of these hard boundaries but has replaced them with a more complex 'networked' border, corresponding to an apparent change in migration flows:

> As the means and the avenues for migrating temporarily across borders increase dramatically – tourism, business, foreign study, family ties – it may not be possible to intercept the 'illegal' immigrant at the border *because they are not 'illegal' at that point.*
>
> (emphasis added, Walters 2002a: 573)

Citing evidence from European Schengen officials, Walters highlights the changing working practices of customs and immigration officials, who now communicate far more with colleagues in other nation states, as well as increasingly focusing on the internal territory of the state (and think of how the speed corridors discussed in Chapter 5 might be fundamentally important new zones of control). Hence the actual policing of borders may involve co-operation with a range of officials, institutions, and departments across and within national territories. This may be in sharp contrast to the UK's island status, where fixed territorial entry points backed up by maritime policing are the norm.

The narrated or symbolic border

While understanding barriers requires attention to both 'natural' features (rivers, coasts), and man-made technologies (passports, police surveillance, barbed wire, walls), both are given meaning by socially constructed understandings. For example, there is nothing to say that the Mediterranean is a 'natural' divide between societies. Indeed, it has been a source of great cultural interaction and understanding, as well as tension and conflict, for centuries. More generally, then, students of borders and barriers need to pay attention to 'narrative and discourse' (Newman and Paasi 1998), in other words the geobabble of daily news coverage, government foreign policy statements, geographical texts (including school maps), and formal and informal myths and teachings and stories about the 'other' and how it gives 'us' identity (think of Germany vs. England at football (Downing 2000), the north versus the south in Italy (Tambini 2001), and prejudiced attitudes to 'non-Europeans' (Pells 1997; Pred 2000)). Furthermore, different social and ethnic groups understand 'boundaries' in different ways. To understand *how* this is the case requires ethnographic study, as there may be strong divergences between how those in the 'frontline' might understand boundaries compared with their colleagues in more remote or metropolitan parts of the same national community.

To return to our example of the Mediterranean, as well as its obvious physicality, it is also intensely mediated. As Driessen (1998) notes, in 1992 and 1993 Spanish politicians and media regularly referred to North African migration in metaphors such as 'invasion', 'flood', 'wave of mass migration', 'being deluged with refugees', and 'Spain under siege from an army of migrants'. Furthermore, he argues that many Spanish or Andalusians often adopt the notion of 'European' identity with greater enthusiasm than, say, Germans or Dutch, as a means of 'marking' Spain out from its underdeveloped past, a similar phenomenon to the 'othering' identified in East and Central Europe (Driessen 1998: 107–8).

Political-technological borders

The geopolitical border should be regarded not merely as a line, a physical location, or even as a symbol, but in terms of a larger heterogeneous assemblage of discursive and nondiscursive practices. Something similar can be said of the biopolitical border. It forms a machine with an assortment of technologies, simple and complex, old and new. These include passports, visas, health certificates, invitation papers, transit passes, identity cards, watchtowers, disembarkation areas, holding zones, laws, regulations, customs and excise officials, medical and immigration authorities.

(Walters 2002a: 572)

Walters (2002a) intriguingly argues that these borders are in fact 'biopolitical' – that is, the border is tied very closely to how *populations* are regulated. At various points of European and world history, huge flows of populations have criss-crossed national territories. In a European context, the First World War and the Great Depression were fundamental events in universalising the use of visas and passports. As a regulatory mechanism, these began to assume huge importance to the establishment of borders, and were linked to ideas of medical screening, the state's assessment of the motivations for migrating, and the relative wealth of various groups. This helped to form ideas of 'desirable' and 'undesirable' populations and migrant groups.

The passport is, therefore, a key technology in regulating the flows of people between different countries. As O'Byrne (2001) suggests, the passport has been transformed during the twentieth century from being a luxury item held primarily by social elites offering protection to citizens when abroad, to a device by which the state could intervene in the life of the individual. As such, the passport is perhaps 'the most important symbol of the *nation-state system*', a 'political tool because it allows an administrative body to discriminate in terms of who can and who cannot travel in its name' (O'Byrne 2001: 403, emphasis in original). The implications of this are clear, as O'Byrne continues: the passport can act as a form of surveillance, as a means of determining who is worthy of belonging to the nation (state) and enjoy its benefits, and – perhaps crucially – determine who has the right to seek permanent employment, start businesses, and so on, within the national economy. And as a further point, it is worth noting the importance of the *forged* passport, perhaps the most sought-after forged document that one can imagine, fundamental to any number of secret service agents or 'illegal' immigrants. The visa – a selective restriction on length or right of entry of specific national groups – offers similarly intriguing insights:

What insights would the genealogy of this little form – at once mundane but also, in its limit cases, a matter of life and death – generate concerning the political and social codification of populations and their movements, the ascription of status, the geopolitical division of the desirable and undesirable, and the regulation not just of the migrant's experience of space but also of their time?

(Walters 2002a: 572)

So, through visas and passports, allied to other techniques such as the keeping of formal statistical records, governments are able to define what is meant by *their* population and decide how they should be governed and regulated.

This summary is only intended to flag up and summarise some of the key issues involved in studying borderlands and barriers to mobility. While this chapter focuses on the territorial spaces where states meet, it is important to consider that they are more than just lines on a map. They are material, concrete realities, policed by 'real' people – border guards, customs officials, soldiers, immigration officers. Keeping in mind our discussion of mobilities, not all borders are on the physical edge of a national territory – they may be in airports hundreds of miles from the actual border posts that we might think of. Furthermore, the impermeability or porosity of borders is determined by technologies, which may be more or less efficient depending on the power and wealth of the state, or the power, wealth, and ingenuity of the unwelcome border crosser (and his or her access to the technology of forged passports, for example).

Box 6   The Berlin Wall

The fluidity and complexity of place identity is in contrast to mythologies of the permanence of borders. Yet even the most 'durable' non-natural border of post-war Europe – the Berlin Wall – only lasted 28 years. Built at the height of Cold War tension by the East German government as a means of stopping the out-migration of thousands of its citizens, the Wall's demolition in 1989 had an extremely complex place within global, European, German, and Berlin geographies. Thrown up almost overnight in 1961, it became an icon of state repression, and a symbol of resistance.

Ladd (1997: Chapter 1) produces a cultural reading of the Wall's significance based around five different understandings: as a monument; as a barrier; as a symbol; as a zipper; and as a relic.

As a *monument*, the Wall was Berlin's leading tourist attraction. Its destruction thus opened up all sorts of meanings about the role of monument. 'Did the concrete lose its aura when it was removed from its original location? Or did that happen earlier, when it lost its power to kill, so to speak – that is, when the guards stepped aside and let the crowds through on November 9, 1989?' (Ladd 1997: 7). And what could be said about the 'wall auctions', where some of the most artistically decorated pieces of Wall were sold around the world in time for Christmas 1989?

As a *barrier*, the Wall divided a previously cohesive city overnight (quite literally) (Borneman 1998). While the West underwent the economic 'miracle' associated with the post-war boom, the East developed at a far slower pace. The closing of the border in 1961 suddenly meant that for Easterners, the whole experience of border crossing became an ordeal. As Garton Ash puts it:

> The Wall was not round the periphery of East Germany, it was at its very centre. And it
> ran through every heart. It was difficult even for people from other East European
> countries to appreciate the full psychological burden it imposed. An East Berlin doctor
> wrote a book describing the real sickness – and of course the suicides – that resulted.
> He called it *The Wall Sickness*.

> (Garton Ash 1990: 65)

Figure 21 *The Berlin Wall (East side gallery) (2002). Donald McNeill.*

The result – shown vividly in Wim Wenders' film *Wings of Desire* – was the creation of curious backwater spaces no longer subject to rational urban planning, in areas that would once have been part of the city centre.

As a *symbol*, the Wall highlighted for many the bizarre logic of Cold War politics. As in the film *Dr Strangelove*, the abstraction of global superpower rivalry was brought to everyday experience in the most absurd ways. While the early years of the Wall were seen as an embarrassment to the West, showing an inability or fear of acting decisively, it was ultimately exploited as the US president's dream destination to make ideological statements. Kennedy's visit in 1963 achieved mythical proportions; Reagan's 1987 appearance 'sounded the metaphor of mobility and connectedness. He stood before the walled-off Brandenburg Gate and demanded, "Mr Gorbachev, open this gate. Mr Gorbachev, tear down this wall"' (Ladd 1997: 22). The deaths of 78 people trying to cross the strip from East to West, and the psychological trauma caused by those whose homes remained next to one of the highest tension fields of urban symbolism ever known, added to the Wall's resonance.

As a *zipper*, (in the words of East Berlin writer Lutz Rathenow) the Wall linked Germans just as it divided them. 'The separation enforced by the Wall made it easy to explain away any apparent disunity among Germans and to render harmless the whole idea of German identity' (Ladd 1997: 30). As such, it was used to narrate the current state of German history and identity.

On the one hand, Germans could interpret official propaganda as implying that the people on the other side of the Wall monopolised the prejudiced, predatory, or authoritarian traits of the bad old days. On the other hand, it was common in both Germanies to characterise the East as the 'old' Germany or the 'real' Germany, implying that the GDR was the repository of traditional German virtues, unspoiled by foreign (especially American) influences (Ladd 1997: 31).

In this way, the Wall helped narrate the troubled nature of German identity. The East presented the West German state as being a continuation of the capitalistic fascism of the Nazis, naming the wall the 'antifascist protective rampart'. The West, of course, reflected on the East German experience as being the 'unnatural' legacy of the historical settlement of post-war Europe. (The concept of the zipper is similar to the debate about the role of the Mediterranean – does it unite or divide?).

Finally, the Wall after 1989 was a *relic*, functionally obsolete, yet historically full of lessons and resonance. What to do with it? As with most contested pieces of landscape, the choice was between memory and forgetting. Most of the latter persuasion saw the retention of the Wall as a painful psychological scar – destruction and the wiping clean of historical memory to move forward was seen as the best approach. By contrast, there was a strong movement in favour of preserving small stretches of the Wall as a historical lesson, either for nationalistic purposes (a symbol of the 'unbreakable unity of the German people' (Ladd 1997: 32)) or else as a manifestation of the inhumanity of abstract political-military impositions on human society.

Cross-border regions

Cross-border or transfrontier regions have been a very substantial topic in European studies in recent years (Kepka and Murphy 2002; O'Dowd 2002). The introduction of a single European market has levelled off many of the differences that separated function-ing economic regions, where trade and commuting has been a long-held trend. Yet as Figure 22 shows, the enlargement of the EU has brought in many countries with differing levels of economic development and infrastructure. As such, transfrontier projects are being seen as an important practical means of integrating post-communist states into the European social and political economy. However, they are also potentially destabilising, certainly of existing geopolitical imaginations, as they:

> expose the chronic lack of fit between nations and states and the arbitrary and coercive
> basis of many borders. They provide a glimpse of alternatives to existing states in
> reforging links across historical, national, ethnic, religious and linguistic entities ruptured by
> the violent process of state formation.
>
> (O'Dowd 2002: 125)

So, the Irish and Basque borders, for example, have often been associated with terrorist activities, and provide important symbolic markers of political division. Similarly, as Paasi

European Borderlands (with case study references)

1. Euregio Meuse-Rhin (Kepka & Murphy 2002, Kramsch 2002)
2. Euregio Neisse-Nisa-Nysa (Kepka & Murphy 2002)
3. Basque borderland (Raento 2002)
4. Catalan borderland (Häkli 2002)
5. Irish border (J. Anderson 1996)
6. Channel Tunnel/Transmanche (Darian-Smith 1999, Sparke 2000)
7. Spanish-Portuguese border (Sidaway 2001)
8. Julian border and Trieste (Kaplan 2002)
9. Trirhena Euregio (Eder & Sandtner 2002)
10. Alto Adige (Kaplan 2002)
11. Øresund bridge
12. Finnish-Russian border (Paasi 2002)
13. Eastern Slavonia (Klemencic & Schofield 2002)
14. Galicia (Bialasiewicz & O'Loughlin 2002)
15. Scandinavian Northern borders (Karppi 2002)
16. The Mediterranean (King 1998)
17. Turkey, Greece and Cyprus (Balkir & Williams 1993)
18. Spain-Morocco (Driessen 1998)

Figure 22 *European borderlands (map).*

(2001) shows, the long Finnish-Soviet/Russian border – and the stories surrounding it – have been an important source of modern Finnish identity.

However, along with the EU's rhetoric of integration and economic co-operation, rather than conflict, cross-border regions have assumed a pioneering role. O'Dowd

(2002: 116–22) discusses three dimensions to this. First, there are the early projects of the post-war period, involving long-contested areas such as the Upper Rhine valley around Basel or the Dutch/German/Belgian border around Aachen, Maastricht, and Liege (Eder and Sandtner 2002). Through project-specific co-operation on issues such as pollution, industrial decline, planning, and cross-border commuting, the basis for a more formal institutional arrangement was laid. Second, the creation of the Single European Market required that cross-border 'stickiness' be tackled, and the European Commission instigated the Interreg funding programme to address the peripherality of these regions in terms of economic decision-making, weak transport and communications links, and the cultural, social and legal/institutional problems associated with differing nation-state regulations. In this logic, once peripheral regions might be seen as 'core' in a new European space. Third, the enlargement of the EU – along with the idea of a new Central European space (or *Mitteleuropa*) – has encouraged the creation of several Euroregions along the German–Polish border, the German–Czech border, the Baltic region, and the Italian–Austrian borders.

However, it is important to keep the nature of cross-border regions in perspective. Interreg draws on a tiny portion of the overall Structural Funds paid out by the EU. Cross-border initiatives may be bogged down due to lack of funding, entrenched political interests, linguistic barriers, and – crucially – the fact that the nation-state is still the major source of policy and strategy affecting their development.

The Channel Tunnel

In 1994, the Channel Tunnel between France and England was completed, a fascinating story which underscores how built structures – whether office blocks, palaces, statues, or engineering projects – can condense and illuminate complex political stories of national or other forms of identity. As Eve Darian-Smith has shown in *Bridging Divides* (1999), conventional narratives of the building of the tunnel centred around the technical issues surrounding the linking of the territories through 20-odd miles of deep water. However:

> In a recent history of the political debates surrounding the building of the Channel Tunnel, a mere ten pages are devoted at the end to the 'intangible and psychological factors' that blocked the building of the link (Wilson 1994). In contrast, I suggest that these factors were the most significant and disabling barriers preventing the Tunnel's realization for almost two hundred years. The engineering and technological debates surrounding the viability of a cross-Channel tunnel or bridge were of considerably less importance than the mental anxieties and competing moral dilemmas that the Tunnel has over time symbolized.
>
> (Darian-Smith 1999: 81–2)

Darian-Smith's analysis is a sophisticated piece of legal anthropology in itself, but her account also reveals the full array of debates surrounding the effective joining of two long-separated territories with a fixed link. The crux of the argument revolves around whether the Tunnel will create a new Anglo-French borderland, or whether these areas will be by-passed by the new high-speed links between Paris and London. According to Sparke (2000), planners and politicians are engaged in 'anticipatory geographies', literally imagining a future territory with clear geopolitical impacts:

> For example, a map of the Channel Tunnel rail services expected to be running by 1996 anticipates the net effect of the new infrastructure in the form of a plan that collapses the complex regional geography of northwestern Europe into a diagram more akin to the maps of the Paris metro and London underground ... At the same time, it is the much wider reach of the service that is highlighted. Stretching from Edinburgh to Amsterdam and from Manchester to Mainz, this infrastructural imagination of the new networks appears eagerly European in scope. *By framing the future in this way, the map is indicative of a wider, infrastructure-oriented geographical imagination that simultaneously shrank the significance and centrality of the borderland space itself* ... More significant than any single map of the new infrastructure linkages has been the network logic unleashed by this mode of reimagining geography.
>
> (Sparke 2000: 198–9, emphasis added)

So, as I suggested in the previous chapter, the development of high-speed, transnational infrastructures threaten to 'tunnel' and 'pump' the cross-border region itself, linking together wealthy and powerful cities at the expense of the borderland. In the specific case of the existing border regions, Kent and Nord-Pas de Calais, Sparke (2000) identifies three overlapping geographical visions: the infrastructural, the Eurocratic, and the entrepreneurial. For Sparke, these visions are present in the national elites that commissioned the Tunnel, the Euroenthusiasts in the European Commission that factored it into their new maps of Europe, and the regional businesses that now market the borderland as a 'competitive', business-friendly space.

Yet along with these visions of the future, the completion of the link also involved a sense of loss, a sense of threat, and a form of nostalgia. Darian-Smith's (1999) anthropology digs deep beneath the identity of Kent, and brings to light how national identities in such 'frontier' regions can in fact be stronger than in the 'core' territory, exploring place psychology and discourse in three ways: first, the idea of the 'other'; second, the role of such borders in revealing deeper national identities; and third, the temporal dimension of spatial boundaries.

First, while the relationship between England and France was at the forefront of the debate, a greater fear existed over the incursion of terrorists, rabies, and illegal immigrants onto English soil, a more generalised English popular anxiety about EU membership:

> The somewhat irrational public fear of rabies can be interpreted as embodying the English people's heightened sensibility of the internal disintegration of their own nation in the face of the encroaching New Europe. At the same time, the rabies fascination sustains the need for the island state's legal defence against external intervention. Thus the rabies scare expresses disillusionment in the establishing of ethnic harmony, which is intimately tied to England's future open borders with mainland Europe. What the image

> of the fast train does is to identify this cultural anxiety and fear. By alluding to England's
> rail heritage, it evokes historical images of the railway's capacity to carve up the
> countryside in the nineteenth century, alter relations of distance between cities, towns
> and villages, and ultimately centralize the industrial nation … In the context of Europe's
> transport network and the penetration of the island nation, the fast train materially
> highlights a turning point in the shifting spatial relations between Brussels and an
> increasingly peripheral England.
>
> (Darian-Smith 1999: 153–4)

Interestingly, and subsequent to Darian-Smith's discussion, the 'storming' of the Tunnel by asylum-seekers held at a French camp (Sangatte) close to the tunnel mouth in 2001 has led to frequent disruption or even closure of the Tunnel's freight services, with high-level diplomatic negotiation ultimately leading to the French closure of the camp.

Second, Darian-Smith intriguingly suggests that the failure of the Kentish (borderland) protest groups to have any bearing on the government's support for the Tunnel reveals a 'feeling that London, the embodiment of the central British legal structure, is failing ordinary people, and that its dominant concerns are now transnational European politics and global economic activities' (1999: 121). Here, the deeply held idea of London at the apex of a national hierarchy with Kent as an idealised countryside (the 'garden of England') is replaced by a cross-channel territorial zone, London becoming more tied into the Brussels–Paris nexus at the core of the EU. Further evidence of this can be found in the Euralille development in Northern France. Here the masterplanner Rem Koolhaas, noted – perhaps slightly provocatively:

> If you imagine not distance as a crucial given but time it takes to get somewhere then
> there is an irregular figure which represents the entire territory that is now less than one
> hour and thirty minutes from Lille. If you add up all the people in this territory, it turns
> out to be 60 million people. So the TGV and the tunnel could fabricate a virtual
> metropolis spread in an irregular manner, of which Lille, now a fairly depressing
> unimportant city, becomes, somehow by accident, completely artificially, the
> headquarters.
>
> (Koolhaas 1996: 334, cited in GUST 1999: 47)

This point is borne out in the enthusiasm with which Kent County Council and Nord-Pas de Calais regional government have acted in the creation of a Euroregion, funded under Interreg, which seeks to co-ordinate areas such as transport, environment, and spatial planning (Church and Reid 1996).

Finally, and as we also saw in the context of the Berlin Wall, borders can be 'temporal', in the sense that they play significant historical roles. Just as East Berlin followed a differing path of development to its western neighbour, with the East 'behind' in terms of the sophistication of its commodities, so the Tunnel was argued to be a temporal intervention. Here, from the Queen's statement at the opening ('To rejoin what nature separated some 40 million years ago has been a recurring dream of statesmen and engineers for several centuries', in *Kent Today* 7 May, 1994, p. 6 quoted in Darian-Smith (1999: 104) through to the actual clues in the landscape, the evolving attitude to neighbouring nations is made explicit:

> This was brought home to me standing atop of Castle Hill (sometimes known as Caesar's Camp), where a Norman defence had once stood. From here, looking down on the Folkestone terminal, you can see a new road cutting sharply into the hill face, underneath grass-covered trenches built against Napoleon's possible invasion and again fortified during World War I. The landscape is deeply layered with historical associations. And this is in part what the English fear about the Tunnel – that their autonomous and distinguishable national consciousness, both spanning and conflating centuries into the present, will be undermined by it.
>
> (Darian-Smith 1999: 104–5)

Thus the Tunnel represents a number of cultural and psychological transformations which deconstruct the fixed essences of region, city, and nation discussed in the first part of the book. As a 'space', the Tunnel exemplifies the tense relation between fast mobilities in the construction of a European identity, and the linking of fixed territory with essentialised notions of belonging, the 'fast' and 'slow' geographies identified by Paasi (2001: 18).

Bridges: Malmo–Øresund, Olivenz(ç)a–Elvas, Mostar

The construction of major infrastructure projects such as bridges is usually underpinned by an economic rationale. But as Sidaway (2001: 744) argues, bridges have a metaphorical or discursive power, 'an encapsulation and representation of a process of "building Europe"', demonstrated by their presence on the back of new Euro banknotes. To consider this point, it is worth briefly mentioning three recent bridging events which have transformed the understanding of national territories and identities.

The first example, which opened in 2000, is the 4.8 mile long road and rail bridge designed to link the Danish capital Copenhagen with Malmo in the South of Sweden, creating a new population of 3.5 million people. However, what is particularly interesting is the decision to 'invent' a new region – Øresund – which seeks to make an economic synergy between two hitherto separate nation states which nonetheless can be 'zipped' together as a 'natural' economic region (Treanor 2000). What is particularly interesting is that this new territory has been consciously branded, in many ways constructing a new cultural and economic space which supersedes the relative peripherality of each side of the bridge from their national capitals (*Washington Times* 1999; Van Ham 2002).

The second example is brought to light in Sidaway's intriguing 2001 study of the rebuilding of a long-ruined bridge on the Portuguese–Spanish border. Here, on one of Europe's longest-established national frontiers, the Interreg programme was used as a means of improving accessibility between two of the EU's least developed regions, Extremadura (in Spain) and Alentejo (in Portugal). The construction of a bridge at Ajuda (inaugurated in 2000) was at first sight a good example of how EU structural funding could be reflected in practical construction projects. What is revealed in Sidaway's narrative, however, is an indication of how almost all border regions have deeply contested histories. The town of Olivenza that lies close to the bridge has long been claimed as being Portuguese; the largely bureaucratic and technical decisions taken over the bridge (and an associated dam) reawakened deep-seated rivalry between Portugal and Spain and stirred up, even in this relatively quiet border region, a deep sense of national sovereignty and historical aware-ness that the transfrontier project had sought to overcome.

My third example comes from Mostar, in Bosnia-Hercegovina. As with Sarajevo, Mostar was one of the symbolic sites of the Balkan war, a city straddling the Neretva River and composed of largely equal Croat and Muslim populations. In 1993, forces from the besieging Croat army shelled and demolished much of the structure of the Stari Most (old bridge), which had been built during the Ottoman empire in 1566. As in Sarajevo, the city had stood as a symbol of largely peaceful co-existence between Croat Catholics, Serbian Orthodox, Bosnian Muslim, and Jewish populations. When set alongside the huge loss of life and displacement in the area it was not the key incident of the war. Yet its symbolism was felt by many, and according to Grodach (2002), the post-war recon-struction of Mostar and its bridge have awoken a series of complex questions about reconciliation, heritage, and cultural identity. Interestingly, Grodach argues that the international news articles and website entries covering the rebuilding process have come to overtake more pressing concerns surrounding post-war society:

> More often than not, these articles contain inaccurate information delivered in a mawkish tone. Such reports not only serve to sensationalize not only the post-war rebuilding process, but also daily life in Mostar. For example, many articles earnestly assert that the front-line of the Bosnian War in Mostar, and current division between East and West (Muslim and Bosnian Croat, respectively) is the Neretva River, once traversed by *Stari Most*, thus lending enhanced symbolic power to the bridge's metaphorical imagery.
> (Grodach 2002: 78)

The rebuilding process reveals the simmering tensions between the resident Croat and Muslim populations who are not as neatly divided as many journalists wish to stress, instead being segregated along a major roadway. Furthermore, while the rebuilding process has been represented by bodies such as UNESCO as a key part of post-war healing, some Croats have claimed that the bridge symbolises a heritage of Muslim (Ottoman) domination.

These three brief examples show in different ways the importance of bridges as a means of condensing, or reawakening, the idea of the border as being constructed discursively. As with the Channel Tunnel, these are more than technical engineering feats. The manner of their construction or destruction, their creation of a new *de facto* territory, means that all over Europe, borderlands are being contested and renewed, and the manner in which this is being carried out is fundamental to any understanding of European territorial identity.

Conclusions

This chapter has sought to focus on borders within the context of a Europe of *mobility*. While in earlier chapters I focused on fixed notions of territory, the last two chapters have suggested that Europe can *only* be fully understood with a complementary understanding of movement. And I have hoped to demonstrate that there is a burgeoning, and very interesting, literature on how these borders and frontiers are being reshaped and lived out in very *human* ways. To conclude, I want to read across some of the themes discussed above to emphasise four points.

The first point is political. The rhetoric of a New Europe as being borderless has been one of the most radical aspects of the integration process. Indeed, there is a strong sense in which national sovereignty – tied into a symbolic national space – is being eroded by the EU. In turn, one of the most important aspects of such sovereignty – borders, and the defence of national space – is being eroded. Yet it is clear that mobility is not being granted as a universal right. Only those who fit within the accepted notions of European citizenship are granted this freedom.

Second, there is a sense in which borders are a complex mix of policing, of discourse and metaphor which reinforce political and social practice, of technology, of administrative systems, of speeches by politicians and racist (or occasionally inclusionary) political parties. Here, major building (or destruction) projects, such as the opening of the Channel Tunnel and the breaching of the Berlin Wall, are often dramatic stages where the full complexity of the insider/outsider distinction – of how states label people as belonging or not belonging – are played out. What is interesting is that the non-human technologies and materials used to build borders and walls fuse with the humans that patrol and police it, and the state officials and politicians that govern or legislate over the space. Taken together, such a combination serves to regulate the mobility of Europeans.

Third, relatedly, this requires a greater focus on the everyday life of borders, bearing in mind that these human-object relations may not in fact be located at the actual frontier of territories. Rather, as Walters (2002a: 577) reminds us, the 'spatiality of the border' requires further research, as the actuality of the border, and its performed nature, may now be found in airports, in the raids on bars and restaurants suspected of employing illegal migrants, in immigration offices tucked away in the back office zones of major cities.

Fourth, how does this affect the idea of the pristine national identity discussed in Chapter 2? Clearly, we can see that if borders ever were wholly constitutive of the nation, the idea of static boundaries which can keep out bearers of 'foreign' ideas are now being challenged by (globally-corporatised) media and cultural flows that are disrespectful of distance. For example:

> Patterns of movement and flows of people, culture, goods and information mean that it is now not so much physical boundaries – the geographical distances, the seas or mountain ranges – that define a community or nation's 'natural limits'. Increasingly we must think in terms of communications and transport networks and of the symbolic

boundaries of language and culture – the 'spaces of transmission' defined by satellite footprints or radio signals – as providing the crucial, and permeable boundaries of our age.

(Morley and Robins 1995: 1)

If this is the case – and there is a danger that this argument can be overplayed – then an examination of the role of broadcasting and cyberspace in Europe is crucial. It still has to be stressed that while my focus here has been on largely corporeal movement (and barriers to it), the effect of flowing images and words – expressive of ideas – is one of the most fundamental challenges to the concept of the national 'border' that there is. As with pollution and global warming, such flows call into question the very nature and validity of borders, given their decreasing strength in 'protecting' the nation from outside influences.

7
Metroworld

> The city is everywhere and in everything. If the urbanized world now is a chain of metropolitan areas connected by places/corridors of communication (airports and airways, stations and railways, parking lots and motorways, teleports and information highways) then what is not the urban? Is it the town, the village, the countryside? Maybe, but only to a limited degree. The footprints of the city are all over these places, in the form of city commuters, tourists, teleworking, the media, and the urbanization of lifestyles. The traditional divide between the city and the countryside has been perforated.
>
> (Amin and Thrift 2002: 1)

While the idea of the metropolis (or indeed the megalopolis) is well-established in North America, Asia, Latin America, China, and so on, in Europe there is still a romantic attachment to the 'Europe of the Cities' idea, that I discussed in Chapter 4. Yet there is a strong sense in which many of the key economic functions, most pressing social issues, most entrenched political identities, are located '*extra muros*' (beyond the city walls), beyond the traditional notion of the compact or dense city. These spaces are not easy to represent, are mundane, boring even, but they are key to any understanding of a mobile geography and sociology of Europe, landscapes caught between fixity and mobility. Airports, our new city gateways, are surrounded by a fixed architecture and urbanity of big sheds and fast transport links. Motorways are engrained in *national* identities as much as in a transnational sense. And any notion of a European citizenship might be faced with the challenge of whether our metrodwellers see themselves as citizens, or as *consumers* of city spaces charged for road use or taxed for new transport links.

To address these issues, I begin by clarifying what I mean by the 'metroworld' in a European sense, placing this in the context of the urban experience around the world. Second, following on from the discussion of driving earlier in the book, I think about the fixed landscapes of motorways, and their associated geographies of big shed retail and distribution. Third, I consider whether these consumer landscapes are altering the nature of citizenship and belonging. Fourth, I provide a reflection on airports and their associated spaces, our gateways, the nodes in our networks, and a metaphor for the fragility of a modern European identity.

Europe's metroworld

> The at times strongly paradoxical nature of posturban space appears most clearly from the countless names that have been devised to denote its confusing constituents: these names range from wordplays on the etymological roots 'urb' and 'burb' (such as slurb, the

> burbs, the technoburb, exurbia, disurbia, superburbia, shock suburbs, surburban
> downtown, suburban activity center, nonplace urban field, dispersed urban regions, the
> rurban fringe) to word combinations with 'city' (edge city, outer city, technocity, galactic
> city, elastic city, polynucleated city, spread city, perimeter city, città autostradale) and other
> labels (sprawl, megalopolis, exopolis, outtown, growth corridor, multinucleated
> metropolitan region, Nowheresville, Anywheresville, autopia . . .).

> (GUST 1999: 27–8)

The starting point in any discussion of Europe's metroworld should address four things. First, that unlike the experience of many other non-European countries, the continent lacks a megalopolis. Following Le Galès (2002, 26; drawn from Moriconi-Ebrard 2000: 315–20), the largest European conurbation – Paris, with 9,850,000 inhabitants – comes 21st in the world (figures based on the year 2000), after Tokyo, New York, Mexico, and Los Angeles sure, but also behind Manila, Moscow, and Istanbul. With London coming 24th, Madrid 48th, Brussels 51st, Barcelona 61st, Manchester 64th, Milan 69th, and Berlin 70th, it is clear that the European metropolis cannot be conceived on the same scale as cities elsewhere in the world. Nonetheless, Kunzmann foresees a scenario of a twenty-first century Europe

> dominated by a huge, fully urbanized Euro-megalopolis consisting of a few cooperating
> global command centres (Paris, London, Brussels, Frankfurt), together with their adjacent
> urban regions (Randstad, Rhein-Ruhr, Berlin, Padania). Efficient intra-metropolitan
> transport and telecommunications systems would link the activity nodes of the
> polycentric Euro-megalopolis to allow easy accessibility to and from international airports
> and comfortable movements within its territory. Nature conservation, outdoor
> recreation, provision of water and production for local consumption would be tasks left
> to the European cities and regions outside the Euro-megalopolis.

> (Kunzmann 1996: 157–8)

Such a vision is broadly confirmed in the projections of the European Spatial Development Perspective discussed in Chapter 5 (Commission for the European Communities 1999), a possible future for an expanding EU territory.

Second, what we think of when we speak about 'Paris' or 'London' is very different from the *metropolitan* area of Paris or London. Yet these, in many ways, are the location of the essentialised 'real' Parisians or Londoners, either driven out by or fleeing from the processes of gentrification or rising land costs discussed in Chapter 4. So Paris has a *city* population of just over 2 million out of nearly 10 million inhabitants, there are 1.5 million residents in the territory of Barcelona city council out of 4 million in the metropolis, and so on. The statistical measurements used to draw these judgements must be treated with caution, but it is clear that the Europe of the Cities, with its cathedrals, grand railway termini, art galleries, and monumental squares, does not do justice to the European metroworld, with its airports, retail parks, and huge swathes of suburban housing. So, there is a need for a fuller consideration of the fragmentation of the term 'city', particularly in terms of the historic 'central places' within a complex network of commuting (by car, train and plane), and goods distribution (Le Galès 2002: 22–5).

A classic idealisation of what a European city might *be* is provided by the Italian urban historian, Leonardo Benevolo:

> The European city came into being with Europe itself. In some sense it begot that region, historically defining European civilization and continuing to be its most salient characteristic as the subcontinent rose to world dominance ... The European city is an integral part of [the modern world] and the preservation of this material heritage is necessary if we are not to lose access to a collection of values not approachable by any other means: the identity of the places in which we live; the stable background against which the flow of diverse experiences that characterise each generation is given significance; the permanence of a 'centre' which changes more slowly than the periphery; a place where we can put together those common memories too weighty to be carried by each individual.
>
> (Benevolo 1993: xv–xvi)

Benevolo's book *The European City* is written from a very traditional conception of urban form, nourished by the historical morphology of the Italian city, and expresses well the desire to preserve or retain the old landscapes of the European city, not only as part of the heritage industry but also as a site of collective memory.

However, European urbanists – from both Italy and elsewhere – are inverting the way in which they look at the city, such as 'to think of the urban from the perspective of "nature" and of the city from the perspective of its periphery' (Keil 1999: 637). So, 'stepping out of the mindset of the medieval city' (ibid. 637) requires us to engage with a European landscape that is post-industrial and consumption-driven, yet which considers urbanity and nature in new ways. This is a European geography of the parking lot, the drainage channel, the airport, and the distribution shed, a far cry from the cathedral, the river, the railway station, and the delicatessen.

In many ways the non-places of the European periphery are perhaps devalued by intellectuals and artists *because* they represent one of Europe's most fearful 'others', America (see Chapter 1). Yet there is an increasing interest across European urban theory in representing these new landscapes from within specific cultural and national contexts. GUST (1999: 32) argue that – compared with the United States – 'a tendency to urban sprawl is no less in evidence in Europe, but the distinction between historic city and periphery remains more firmly in place'. This distinction – an on-going fascination with the spaces of cathedral and monument, of park and square – remains fundamental to the image of the European city. Yet there is a growing quantity of work that actively problematises the idea of the European periphery and tries to explore and reimagine it.

Third, Europe's economic landscape is changing rapidly. We have technopoles, for example. Following the classic example of Silicon Valley, the likes of Cambridge, Sophia-Antopolis (Nice) are centres of higher education and research. To develop Keil's point made above, universities are now seen as being crucial generators of high technology research, as seen in the classic US examples of Silicon Valley and Route 128. For Miyoshi (2002: 61), these 'are the late twentieth-century campus landscapes that have replaced the Gothic towers of Heidelberg with their duels, songs and romance, or Oxford and Cambridge with their chapels, pubs, and booksellers'. As I shall discuss below, major airports are now also seen as highly profitable centres for a variety of business functions, including conference facilities, retail outlets, and distribution hubs. The European motorway network has brought with it the urbanisation of, for example, the Rhine, Rhone, and Danube valleys in Germany, France, and Austria respectively, giving us inter-regional distribution centres and corridors. Due to the congestion of urban cores, a whole kaleidoscope of modern urban functions which relate to the moving of goods can now be found in the urban periphery (Baart *et al* 2000; Neutelings 1994). In extreme, it could be suggested that if urban sprawl continues the European continent may come to resemble the US with its edge cities, where commuting may be *between* edges, rather than into the historic city core (Kunzmann 1996).

Fourth, there is the idea of homogenisation and 'non-place':

> The real rhythm of local life, away from bright lights and shop-windows, shows itself in suburbs. And what first appears is the likeness of one to another. Staying in suburbs of Hamburg, Brussels or Bonn, I found it often hard to remember that this was not England or America ... the same small family units, with their car in the garage ... the television set (sometimes showing the same English programmes); ... the same little status-symbols, garden objects, gnomes, old lamps, fancy letters on the gate; the same kind of supermarket at the end of the street, selling Daz, Ajax and Persil. Even the smells in the suburbs are the same: the unmistakable city smells of local food, sanitation or sweat give way to the supranational soaps, antiseptics and petrol fumes.
>
> (Sampson 1968: 215)

There is a strong argument within the social sciences that more advanced forms of capitalism experienced in the developed world are erasing local cultures, authentic experience, and so on. In particular, franchise and chain forms of retailing – be it motorway service stations, Gap stores, Barclay's Bank, McDonald's, or Marriott hotels – are rendering the continent homogeneous. The economic landscapes discussed above are designed in such a way as to encourage the quick processing of humans and things, to reduce elements of unpredictability. These spaces are arguably Europe's lifeblood, it is, like it or not, where the operations are located needed to secure the standard of living to which Europeans aspire. This is perhaps because they are, in the terminology of French anthropologist Marc Augé (1995), 'non-places'. By this he means that 'capitalist modernity creates a distinct mode of mundane locational experience' (Tomlinson 1999: 109). On the one hand, then, Europeans are increasingly living in suburbia, away from the deindustrialised cores. On the other, their *experience* away from their homes is thus characterised by a geography of movement, of non-permanence, of mental mapping that does not respect measured, statistically mapped space.

Box 7   Maspero's *Roissy Express*

You have to keep repeating it: there is nothing remotely geographical about the place. It's simply juxtaposed horizontal and vertical divisions which are impossible to take in immediately: between the connecting roads cutting under and over artificial embankments, sometimes with long bends well over 180 degrees, almost circular even – a swing left, a swing right, and you always see the sun just where you're not expecting it – between buildings rearing up here and there and blocking out the view, barely identifiable and almost anonymous cubes and towers, which at first glance, at least, are no use as reliable markers, then there's all the asphalt running over your head, the railway line, and the motorways you keep coming across, the bridges and tunnels, and all the vehicles pelting along, overtaking, overlapping and separating, watch your left, watch your right, and not a single pedestrian to give the whole thing a scale … But who's asking you to make sense of something that is only for travelling through? And quickly. By car. Even if it means going round and round and round. These are temporary spaces.

(Maspero 1994: 20)

They're perspiring, they're hungry, they talk about the sandwiches waiting for them in the café in Villepinte. Larks fly away from the green wheatfields towards the sun, and their strident song grows fainter. They disturb a partridge and find rabbit droppings among the broom, the blue flowers, the buttercups, and the puffball dandelions with which, blowing softly, you can play 'She loves me, she loves me not … '. Down below, towards the main road, a sign announces a forthcoming 'caravan site for people of the road' marked by a concrete surface. It would be hard to find anywhere more remote than this godforsaken place.

(Maspero 1994: 59)

Two contrasting quotations, two contrasting landscapes, but both drawn from the periphery of one city: Paris. Whether it be the approaches to Roissy airport or the semi-rural backwaters of Villepinte, the diversity of spaces a matter of kilometres from major cities is worthy of greater reflection than often afforded by urbanists. In 1989, the French anthropologist François Maspero set out with his photographer companion Anaïk Frantz to explore the Parisian periphery, using the suburban rail network, the RER, as a structure. *Roissy Express* is a wonderful and thought-provoking read, and perhaps it is so because of the author's awareness that he is visualising the Parisian periphery as if visiting an exotic land:

You poor fool, you want to tell other people about other people's worlds, but you can't even be bothered to tell yourself about your own world – you can always look competent and professional when announcing that people in Shanghai have two metres square living space each, but what do you know about the way people live half an hour from the towers of Notre Dame?

(Maspero 1994: 7)

The trip was conceived as a journey, just as any anthropologist would undertake:

> For a month, then, they would go far away from home, saying goodbye to their families, as you would when setting off for any country you want to visit. He would make notes, she would take photographs. It would be a nonchalant sort of stroll, not an inquiry; they had absolutely no intention of seeing, explaining and understanding everything. The basic rule ... was to take the RER from station to station and, each time, to stop, find accommodation and have a walk round. They would look at the landscapes, admire or detest them as the case might be, search for traces of the past, visit museums and go to a show if the opportunity arose; they would try to grasp the geography of the places and the people – to see their faces. Who were the people who had lived there? How had they lived, loved, worked and suffered? Who lived their today?
>
> (Maspero 1994: 13)

This mode of urban exploration takes the travellers through a landscape at times surprising, at times grey and bland, but above all – and despite my choice of quotations above – thoroughly 'peopled'. In many ways it is a tribute to the method adopted by Maspero, a style of 'thick description' that captures every detail of the landscapes they traverse. To the appalled amusement of the travellers, upon publication the book immediately entered public debate as if throwing completely new light on the territories only minutes away from the centre of Paris, which perhaps testifies to the power of evocative urban writing, but also to the lack of interest shown by city dwellers to the periphery.

However, there is potential pitfall involved in representing the periphery, a problem confronted (but not always recognised) by urban ethnographers since the birth of the discipline. This is the search for 'local colour', the exotic, the danger of writing about these exurban communities from a perspective of metropolitan snobbery. New Yorkers may have disparagingly invented the 'bridge and tunnel' put-down to castigate the weekly colonisation of their favourite bars and clubs by up-state revellers, and hostility to the suburbs is alive and well among many intellectuals:

> The well-to-do liked our book because it was reassuring. One of the standard types of discourse on the suburbs is ultimately no more than a continuation of the nineteenth-century line on the lower depths and the dangerous classes. Eugène Sue's Mysteries of Paris have been replaced by the mysteries of the suburbs. We came along saying that they weren't as mysterious as all that, and people were relieved; they could breathe again.
>
> (Maspero 1994: 264)

Yet there remains a lingering danger of a morbid fascination that fiction or cinema audiences have for dystopian places (a form of 'dark tourism'). Perhaps the new periphery might be the location for this. As Wilson puts it:

These, not the slums nor the sewers, are the new site of the *mystères de Paris*. This is where the white trash culture, which so fascinates Hollywood, belongs. It is the new heart of darkness, a city of dreadful day rather than dreadful night – atrocious crimes carried out in the glare of Californian sunlight or in the bleak boredom of Aubervilliers or Drancy, places even more thrillingly alienated than the dark Victorian slums, which, by comparison have come to seem quite homely and familiar.

(Wilson 1995: 160)

Regardless, *Roissy Express* is a classic of urban writing. Just as he unlocks the Parisian periphery for befuddled intellectuals, so his method could be put to good use in exurban Rome, Frankfurt or Madrid.

Roadscapes

Metaphors and similes. The M25 is nothing in itself. It is 'like' a crop-circle, a doughnut, a tagging device for delinquents, a stun fence. A river. London Orbital has replaced the Thames in our mythology. It transports the dirty cargoes of the city: toxic waste, megastore re-stocks, cattle to slaughter, carcinogenic contraband and desperate economic immigrants.

(Sinclair 2002a: 18)

At various points in this book I have discussed the importance of the car and rapid transport to European territory, in terms of trans-European mobility and national car cultures respectively. If we add to this the link between driving and shopping, and advertising and television watching, we could plausibly suggest that in Europe – as in the US and elsewhere – society is organised *around* the car. Thus images of the role of cars in allowing the 'good life', again related to the prevalence of advertising in supporting this culture of mobility can be played off against the extent to which cars and roads denigrate the natural environment, from noise to air quality to resource extraction to destruction of wildlife habitats for road-building (Sheller and Urry 2000). While the EU has tended to favour more fuel-efficient cars than the US, it will soon have to reconcile the promise of ever-increasing living standards (on which it depends to justify its existence) with ever-decreasing natural resources on which to sustain the long booms to which many Western Europeans have been accustomed, and which citizens of the new member states will expect.

Within this expanding market, such consumer demands have a major impact on several key economic sectors, such as retailing and distribution. The latter activity, essential to any advanced market society, has brought its own distinctive architecture to the European periphery, what Martin Pawley (1998) identifies as 'Big Shed' architecture:

> 'Big Shed' architecture ... began in earnest in the early 1970s when all over Europe, in a great dorsal belt running from the English Midlands to the Mezzo Giorno, the new distribution landscape of the European Community first began to come together. In place of traditional town and city locations, giant mechanised distribution-centre floorspace began to be constructed at breakneck speed at thousands of exits and intersections on nearly 50,000 kilometres of autoroute. During the 1980s a million out-of-town commercial and retail centres sprang up to join them, with no reference to the fate of the ancient town and city centre sites left behind ... Planners and architects played only a minimal role in the production and operation of these 'zero defect' enclosures, devoid as they were of any art-historical identity. These blank-walled buildings were visible manifestations of the abstract, invisible, digital network that now links all the EC countries and their neighbours in a seamless web of production, distribution and consumption ... at present this 'digital urbanism' (its 'towns' are often only designated by numbered autoroute exits) is culturally ignored. Yet in economic terms it is already of far greater importance than anything built inside the old towns and cities it has bypassed ... Unlike heritage architecture, which has the vast literature of tourism to support it, this is ... 'undocumented construction'. There is no cultural literature to document it. No novelist or film maker explores beneath its surface. Who in the arts knows anything about the culture of truck drivers who sleep in tiny capsules above the cabs of their trucks, their positions plotted and checked by satellite ... Who chronicles the doings of fork-lift truck drivers, checkout persons, air traffic controllers, mechanics, linemen, canteen operatives, cash card loaders, vending machine loaders, photocopier repairers, stacking crane drivers or security guards? These are the prototype non-communal persons of the future, denizens of the 'Big Shed' universe, linked to one another only by the global heartbeat of FM radio and satellite TV.
>
> (Pawley 1998: 193–5).

Note well: the great products of the humanities of university life played no part in creating this landscape. No architects, no civic designers, no poets here. Only the economist, the management theorist, and the engineer.

These new building forms and service workers are part of a different aesthetic to the medieval or imperial building legacies visible in major European cities. They form part of what Bell (2001) sees as a shift from architecture to 'carchitecture', as the daily experience of most Europeans may have as much to do with slip-roads, traffic lights, car parks, speed humps, and traffic cones. For Sheller and Urry, the car forms part of:

> an extraordinarily powerful *machine complex* constituted through the car's technical linkages with other industries, including car parts and accessories, petrol refining and distribution, road-building and maintenance, hotels, roadside service areas and motels, car sales and repair workshops, suburban house building, new retailing and leisure complexes, advertising and marketing, urban design and planning.
>
> (Sheller and Urry 2000: 738–9)

Some of the implications of this I discussed above, with the suggestion that the nuts and bolts of European society lie not in its cathedrals or squares, but rather more in its transport infrastructure, fully understood.

As a retort, there is a growing interest in placing these peripheral zones at the centre of our imagined European geographies. There are many imaginative engagements with these spaces. For Dutch architect Willem-Jan Neutelings (1988, 1994), the 'ring zone' or orbital landscape of cities has been given insufficient attention, as GUST summarises:

> While architects and urban planners have heated debate on stylistic issues and cosmetic adaptations, the largest part of building assignments relates to peripheral areas, where they are solely determined by economic and bureaucratic powers. The ring zone is an architectural free zone, which shirks any planning and turns a blind eye to the supervision that is deemed so important in the historic centers. Thus, today's planners repeat the mistakes of the modern movement, which slighted the historic cities and in so doing gave free play to real estate speculation.
>
> (GUST 1999: 37)

From the UK, Patrick Keiller's (1999) film and book *Robinson in Space* seeks to penetrate beneath the contemporary English landscape, meshing its cultural, economic, and political geographies through a series of striking, still images. He provides symbolic, stark postcards: a loaded container vessel at Felixstowe, a floating reminder of the UK's balance of payments problem; the half-glimpsed hulk of an Eddie Stobart articulated lorry; a road-sign arrowing to 'Samsung' or 'Toyota' as if these corrugated plants were towns with churches and pubs; a snap of a Warner Brothers multiplex cinema at Lakeside, the UK's largest mall in terms of shops. In its book form, Keiller's visualisation of the new England includes an eclectic series of musings, interjections, and observations, including a discussion of the respective merits of the motorway hotel chain, between the available options of a Travel Inn, a Forte Travelodge ('generally smelt of air freshener and had the charm of being slightly dated', p. 214), or the Granada Lodge; the reminder of Warrington's iconic status in British cultural studies as being the location of its first ever Ikea; loving still colour photographs of British power stations and waste disposal depots, those unsung heroes of our unsustainable ways of life; a meditation on World No-Golf Day (p. 14), prompting thoughts on the golf course as a space-guzzling threat to the peri-urban; and the continual revisiting of the British out-of-centre supermarket: 'With fountains worthy of Versailles, Dorchester's was the most exotic Tesco encountered during seven months of travelling' (p. 94). It is difficult to fully evoke *Robinson*. But as a condensation of the visual codes of the changing British economy, it describes a landscape of mobility and transformation that is so difficult to render in orthodox representations.

In 2002, Iain Sinclair published *London Orbital* (Sinclair 2002b). As with Maspero's rail-based odyssey, this was a walking tour structured around rapid transit infrastructure, a journey through a London far from Trafalgar Square and Notting Hill. He offers up the M25 as a hidden shrine to how *infrastructures of mobility* define a very political version of national pride, in as eloquent a way as the iconic landscapes described in Chapter 2: 'the M25 (Old Tory) was the precise contrary of New Labour's Millennium Dome on Greenwich Peninsula: it was much, much more popular than it was supposed to be'

(Sinclair 2002a: 16). 'Why the M25? Why walk around it? That was my project (a number of soothing days out in the air, making a leisurely circuit). I had to find some way to exorcise the shame of the Millennium Dome ... To find out where London gave up the ghost'. (Sinclair 2002a: 17–18). Sinclair's focus on the motorway raises various issues concerning representation, particularly the idea of *walking* around a space that is characterised by *speed*. Such a strategy works against the fact that motorway landscapes effectively disengage the user from the city, and raises the problem of how to *map* places characterised by rapid mobility:

> Classic maps are simply too static to capture forms of territorial change linked to movement. For example, many of the signs and advertising hoardings which dominate the peri-urban landscape are designed to be seen on the move. Motorway service stations – visited by 200 million people at least once in Italy every year – are absent from cartography, yet represent a key and familiar part of the urban landscape – the new urban *piazze*.
>
> (Foot 2000: 18)

There is a growing literature on the metroworld as being dominated as much by speed and time as by location in relation to the historic centre:

> Architecture, planning and urbanism often still remain tied, however implicitly, to the classical analytical tools of the perspectival plan, or the formal geometric composition, even though awareness of the failings of such paradigms for representing urban life is increasingly widespread.
>
> (Graham and Marvin 2001: 412)

What really characterises this new model of the metropolis is the relationship between space and *time*. The development of high speed corridors – either motorways or rail networks – combined with an increasingly complex air network, with an associated rise in specialised 'airport towns', has transformed the logic of core and periphery. And, furthermore, this has very clear implications for how the city or metropolis is imagined. This problem is familiar to those who know American cities, and who find the metroworld *disorienting*:

> Traditional-downtown urbanists recoil because a place blown out to automobile scale is not what they think of as a 'city'. They find the swirl of functions intimidating, confusing, maddening. Why are all these tall office buildings so far apart? Why are they juxtaposed, apparently higgledy-piggledy, among the malls and strip shopping centers and fast-food joints and self-service gas stations? Both literally and metaphorically, these urbanites always get lost.
>
> (Garreau 1991: 9)

So, mobilities are completely disrupting our stable maps of Europe, and in truth they always have, at least since the speed-up of transport forms which intensified in the industrial revolution. But how do they fit in our basic geographical knowledge? We know the capital cities, but we don't know their suburbs. We know Amsterdam, but may only meet the Randstad at higher levels of school or university. We know Milan and Florence and Bologna, but the idea of a Padana metropolis where these medieval cores and their peripheries are linked by motorway is not in our collective consciousness. There are a number of skills that Europeanists and urbanists require to become comfortable with discussing this new landscape.

So, unlike the properties of classical architecture and planning – with monuments, squares, and carefully considered perspectives – there is a new mode of appreciating the metroworld, a fragmentary, even cinematic, experience:

> In the organisation of space around the strip, the simultaneity of visual stimuli, which dominated the image of the modern metropolis, has vanished. Visual coincidence has been replaced by a gradual concatenation of images. As the posturban landscape is the realm of motorized traffic, the speed at which these seriated signs and stimuli are read depends on how hard one chooses to step on the accelerator. The chaotic spectacle of the modern metropolis and the modernist principle of zoning, which was precisely out to neutralize and counter that chaos, appear to merge again in the posturban landscape ... The increasingly long distance between different urban fragments is reduced by the car, yet that car has also turned the spectacle of the modern metropolis into a cinematic experience.
>
> (GUST 1999: 45)

In Italy – usually noted for an urbanity based on fountains, *piazze*, colonnades, street cafes, and statues – the whole idea of what Italians call *la città diffusa* (the spread-out city) is now the dominant mode of everyday life for most Italians. In common with the words-image combination of most of the works discussed here, the Italian photographer Gabriele Basilico and sociologist Stefano Boeri produced a fascinating, if superficial, discussion of the changing landscape of urban Italy. The vocabulary they use is familiar from the sociologies of mobility discussed earlier: they reflect on the 'city of distant relationships', choosing six rectangles of land, 50km × 12km, which correspond to Italy's 'ribbon' development – Milan to Como, or Naples to Caserta, or Venice to Treviso, for example (Basilico and Boeri 1999). They are motivated by a frustration with 'topo-graphical maps', starting instead with the mappings of density offered by satellite photo-graphy, but then proceeding with an intention of constructing 'other kinds of maps: maps capable of registering the rhythm of life in these places, maps of movement, of flux, of sequential perception, maps capable of depicting the cycle of spatial and temporal life and of discovering the multiple 'populations' inhabiting one metropolitan area' (Boeri 1999: 12). This – an 'eclectic atlas' – shares with *Robinson in Space* an intense interest in how the socio-economic culture of the country is reflected in the landscape, a testimony to how Italy has been transformed over the last few decades, with the poignantly empty images of a cheap, privatised landscape portraying a country with little interest in the public spaces and architectures for which it remains so justly famous.

To capture this complexity, imaginative techniques of writing, mapping and other forms of representation are thus required, whether this be the eclectic atlas of Boeri and Basilico, the anthropologically flavoured travelogue of Maspero, or the cinematic stills and laconic commentary of Keiller. Yet above all it suggests a shift of attention away from the dense, image-rich chocolate box of our urban cores, the bombast of our national identities, and greater interest in the mundane economic landscapes of our artless urbanity. J.G. Ballard summarises it well in the context of the UK:

> What happens between the M3 and the M4 will define the character of Britain for the next half century. American writers, painters and film-makers set off for Los Angeles 50 years ago, abandoning Chicago and New York, because they guessed that Southern California's amorphous sprawl contained the key to America's future. So, abandon that Spitalfields loft and head out for that virtual city waiting for its Edward Hopper and Ed Ruscha, its Rimbaud of the video rental store, its Warhol of the shopping malls. As always, the most exciting spaces of the imagination are where you least expect to find them.
>
> (Ballard 2001)

An embrace of the ethnographic methods and travel genres as used by Maspero or Sinclair? The photos of Basilico? The eclectic atlas of Europe has to be upon us. To repeat Ballard: 'the most exciting spaces of the imagination are where you least expect to find them'.

Consumer landscapes

> For the sad thing about almost a century of history of the modern city and of modern urbanization is that none of the recent outcomes of that history any longer meet with approval. The classic modern ideals of metropolitanism or cosmopolitanism have lost their glitter and the urbanite as a world-citizen has been superseded by the *commuter*, the type of urban dweller who does not want to be an urbanite ... The suburban way of life has become a crucial part of the identity of late-modern citizens. It marks their views of the good life, determines their self-esteem and individuality, exerts a major influence on their roles in public life, their political beliefs, their cultural sensibilities, and their attitudes towards others.
>
> (Boomkens 1999: 219–20)

The view that modern versions of citizenship – based on mutual dwelling in a tight-knit community, for example – have broken down has gained considerable credence. But is it overplayed? Are we now seeing urban dwellers who wish to keep their contact with other urban dwellers to a minimum? Probably, but is this a new phenomenon? And how do we know as social scientists if this is the case? In Chapter 1 for a while I spoke about the 'othering' process going on in Europe, which included internal 'others', there is a very clear sense that *urbanite* academics may have an inbuilt hostility to those who commute or live in the suburbs.

Traditionally, the growth of mass transit and suburban living has given rise to the 'spiderweb' form of urban development, in which major routeways in turn stimulate industrial and retail relocation, and provide the accessibility that commuters have demanded to get to their jobs in the old urban core. More recently, in a process certainly visible in North America but also, crucially, in many European areas, the whole commuting logic has changed. People may work, live, and play around the edge of the metropolis, finding no need to travel into the core at all (see Keil and Ronneberger 1994 on Frankfurt, for example). This has led urbanists to speak of the 'network metropolis', in which the 'dichotomy between center and periphery has lost much of its validity ... since the interurban periphery has acquired a certain autonomy ... In some smaller cities or towns, the periphery has itself even begun to function as a magnet (through its shopping malls, movie multiplexes, and so forth)' (GUST 1999: 35). In many ways this is the European version of the 'edge city' (Garreau 1991), where privatised landscapes of shopping or residence reflect a turn away from the modernist city centre, and where the sprawls of two or more cities almost join up to form a new, 'edge' place.

So where does this leave citizenship and civic responsibility? The traditional idea of the agora or the public square in the heart of cities as the site of political identity may now be superseded by a new form of citizenship – that of consumers (who can now vote in supermarkets, after all). As Sheller and Urry (2000) note, cars lead to the dominant form of '"quasi-private" *mobility* that subordinates other "public" mobilities of walking, cycling, travelling by rail and so on', as well as dominating how people organise their work, family and leisure life (Sheller and Urry 2000: 739). Again, the extent to which this is atomising European society, encouraging a culture of individualism, is open to debate. When comparing the North American experience of space and sprawl, with the more confined nature of European territory, policy-makers puzzle over the extent to which taxation, land use, and business location should encourage or discourage increasing car use. Across European national cultures there are substantially differing public opinions on how far this should be followed.

One extreme example can be found in Italy, near Milan. Here, in the new town Milano 2, built in the 1960s by Italy's prime minister-to-be, Silvio Berlusconi, Foot (2001) finds probably the most advanced example of the dreaded television-privatised suburb nexus that now runs to the heart of Italian politics:

> Above all, Milano 2 epitomised a new model of consumption – a 'city of number ones' as the slogan went. The whole project was built so as to enclose the residents within a model of wealth, a non-urban environment and space. Milano 2 therefore was a complete way of life, a status symbol, and not just an ordinary housing project. Berlusconi made sure that the residents were isolated from the nasty aspects of urban life – traffic, crime, immigrants, workers, the city itself ... One of the services included with a house at Milano 2 was private cable television.
>
> (Foot 2001: 100–01)

Berlusconi used this local channel as a means to propel himself into the Italian television market, which he now dominates. Intriguingly, it was his broadcast of the likes of Dallas that guaranteed that success, part of a diet of American-style quiz shows, soaps, and sport. And as we saw in Chapter 2, his ascension to the post of Italian prime minister was based upon the adoption of American political marketing strategies. Milano 2, then, stands as a microcosm of how a national culture will develop its own, peculiar form of 'Americanised' living.

To take another example, the transition from social democracy to a more neo-liberal economic model has had a significant impact on Swedish society:

> [T]he Stockholm region in the 1990s shows an extraordinary division between two contrasted landscapes, so different in character that they might have developed on either side of an ideological wall – which, in a sense, they did. On the one side is the Social Democratic landscape of the 1950s and 1960s: uniform, standardized, in the best possible taste, embodying the powerful ideals that brought it into being. On the other side is a landscape of the 1980s and 1990s: a placeless landscape that might have come out of New Jersey or Texas, Americanized, blatantly commercial, celebrating the collapse of the Social Democratic consensus. [To the North of Stockholm] the visitor passes from Kista – the last of the great satellite developments, completed only at the end of the 1970s – into another world, the creation of the 1980s and 1990s: a vast linear Edge City of business parks and hotels and out-of-town shopping centres, stretching along the E4 highway, for twelve miles and more towards the Arlanda airport. It is almost indistinguishable from its counterparts in California or Texas.
>
> (Hall 1999: 878)

This description of the Stockholm periphery captures a national cityscape in political transition. Perhaps the most determined attempt since 1920s Vienna to engineer a metropolis on the basis of communitarian values, funded by an economic policy based on high marginal rates of taxation, the 'Swedish model' – which had sustained the hopes of many on the Left throughout the 1970s and 1980s – has been undermined. The satellite new towns on the edge of the capital, such as Vällingby and Farsta, which were begun in the 1950s, were designed with strict attempts to marry home, work, and community with land use rationally linked to a system of public transport. As Hall (1999: 882) puts it, 'there was a distinct and quite conscious set of biases in SDP [the ruling Swedish social democratic party] housing and planning policies: they were designed to produce a certain kind of society and then to reinforce it'. Through mass, high density, housing, efficient public transport and collective facilities such as laundries and play-grounds, there was a lurking ideological attempt to homogenise and guarantee the SDP's hegemony on Swedish national politics and identity.

These two examples draw attention to the impact of American urban form on distinct national societies, and – given Berlusconi's dramatic extension of this model into electoral politics – raises questions about citizenship and belonging. It is increasingly common to hear reference to a *consumer citizenship,* where orthodox forms of identity (loyalty to the nation state) are fragmented by any number of class, corporate, cosmopolitan, sexual, diasporic, or cultural forms of citizenship (Isin and Wood 1999). In this context, often referred to as 'postmodern', an increasingly individualised culture replaces one based around loyalty to the nation-state.

> The postmodern culture thesis interprets the shift from manufacturing to service industries, the increasing importance of design, packaging and advertising in the production and consumption of commodities, the proliferation of an infusion of commodities into everyday lives ...The postmodern culture is then infused, on the one hand, with the circulation of non-material commodities ranging from advertising, information, databases to personnel and client relations management, producer services, planning and marketing functions and, on the other hand, with less labour being spent on transforming matter and more on imagineering. Business corporations now fancy themselves as purchasing not labour but commitment, personality, emotional warmth, personableness and sincerity, and selling not products but the fulfilment of desires, commitment, image and lifestyle.
>
> (Isin and Wood 1999: 149)

In this context, the landscapes of the European periphery will often reflect aspects of this culture, from advertising billboards to huge furniture stores, car showrooms to convention centres. The 'neo-liberal landscape' that fits with Milano 2 or the Stockholm region will include privileged access to computer technology and Internet infrastructure, shopping malls, gated communities, and, perhaps, privatised toll roads. And the huge distribution nexus alluded to in the discussion of ports in Chapter 5 highlight the largely invisible, but extremely significant, movement of consumer goods, continuing with Keiller's (2001) identification of the UK road system as a 'publicly funded warehouse':

> This is most obvious on summer evenings, when busy trunk roads on which parking is permitted become truck dormitories: south of Derby, an eighteen-mile stretch of the A42 lined with lay-bys that connects the M42 with the M1, is one of these; the nine-mile stretch of the A34 between Oxford and the M40 is another. Many of these trucks are bound for the enormous warehouses of inland distribution estates near motorway junctions – *Wakefield 41*, for example, at junction 41 of the M1, next to its junction with the M62. The road haulage – or *logistics* – industry does not typically base its depots in port cities, though it is intimately linked to them: the road construction battlefields of Twyford Down and Newbury were the last obstacles to rapid road access to the port of Southampton from London ...The relative insubstantiality of industrial development in the modern landscape seems to be accompanied by very high levels of energy consumption.
>
> (Keiller 2001: 448–9)

Elsewhere, Keiller notes that the Midpoint distribution centre in Cheshire, owned by the Tesco supermarket chain, has a floorspace of '375,000 square feet, ... 33,000 pallet positions, and a holding capacity of 43 million cases, or nine days' stock for 120 stores' (2001: 450).

So, Europe's periphery is in many ways the place where its economic activities, and consumption choices, are facilitated. The freedom of movement of goods, central to the

Figure 23 *The port of Hamburg. © Audiovisual Library European Commission.*

Single European Market, is actioned here. Furthermore, the freedom of movement of people, and the fulfilment of individual consumption desires, is enshrined in a consumer citizen's 'rights' to travel, as I now explore.

Airports

> The linear border enclosing its national territory is a historical, not an eternal
> phenomenon. Given the significance and centrality of air travel to human migration today,
> the space of the border is now shaped powerfully around the international airport.
> Hence, further research needs to trace this reconfiguring of the border from a space of
> lines and edges to one of nodes. How do airports in EU countries represent places where
> 'external' borders are now on the 'inside'? How might we read the airport as a strategic
> locale where administrative practice seeks to reconcile liberty and security? How is this
> airport border being dramatized under the terms of the 'war on terrorism'? These types
> of questions should be at the heart of any future attempt to map Schengenland.
>
> (Walters 2002a: 577)

Airports – defined by David Pascoe as '"vessels of conception" for the societies passing through them' (2001: 10) are a fundamental part of the European metroworld, socially, culturally, and economically. Whether in their role as a continuation of borderlands, or in their perhaps subconscious marker of home and travel, or even of cultural specificity in an apparently bland non-place, the airport has multiple functions in an integrating Europe. I identify four.

First, airports are *social* spaces, clearly defined territories of cultural mixing. Deyan Sudjic (1992) describes Heathrow as of equivalent importance to, for example, a major public square:

> Along with a handful of other international airports, Heathrow has reached the critical
> mass that makes it much more than simply an interchange between passengers, aircraft
> and cargo … Despite the fact that large areas of the airport are restricted to people
> who have tickets, it has authentically urban qualities that self-consciously fabricated tourist
> traps such as New York's South Street Seaport or London's Covent Garden do not. It is
> a place which attracts tourists and plane spotters, job hunters and salesmen, criminals,
> retailers and caterers. Businessmen come here for conferences. Punjabi women from
> nearby Southall find work as cleaners and kitchen staff. Entrepreneurs, fresh from their
> business school MBA courses, come to put their marketing theories into practice with
> new ideas for retail franchise chains. Its immigration halls have become the setting for
> political demonstrations. Tamil refugees have stripped naked here rather than quietly
> submit to deportation. Libyans have bombed the baggage hall and, along with those at
> Gatwick, the airport's approach roads are the only place in mainland Britain on which the
> army regularly deploys armoured vehicles. The Wapping paparazzi keep the place
> permanently staked out on watch for passing celebrities. Clearly this is as highly charged
> a part of the public realm as Trafalgar Square. By most reasonable definitions it is as
> urban as you can get, a forum as well as a gateway.
>
> (Sudjic 1992: 152)

The crucial – if polemical – argument made here is as follows: that airports are for Sudjic
the 'new city square', as central to our public mixing as any other urban space. And, 'The
poor and the very rich come into closer contact at an airport than almost anywhere else,
despite the elaborate hierarchy of classes and facilities for them' (1992: 153).

Second, paradoxically, airports are *postmetropolitan spaces* (cf. Soja 2000), in the sense
that they both decentre the historic core and – if unchecked – present a logic of
unplanned urban sprawl. In this sense they fit with the argument of Marc Augé, that these
spaces are designed for rapid, non-social movement, centred on individual mobility:

> Place and non-place are rather like opposed polarities: the first is never completely
> erased, the second never totally completed; they are like palimpsests on which the
> scrambled game of identity and relations is ceaselessly rewritten. But non-places are the
> real measure of our time; one that could be quantified – with the aid of a few
> conversions between area, volume and distance – by totalling all the air, rail and
> motorway routes, the mobile cabins called 'means of transport' (aircraft, trains and road
> vehicles), the airports and railway stations, hotel chains, leisure parks, large retail outlets,
> and finally the complex skein of cables and wireless networks that mobilize
> extraterrestrial space for the purposes of a communication so peculiar that it often puts
> the individual in contact only with another image of himself.
>
> (Augé 1995: 79)

Yet these are by no means disconnected spaces. As Sudjic (1992) notes, they will be
major employers of local communities. Major airports – Schiphol, Heathrow, Malpensa,
Barajas – have a substantial service sector workforce, in catering, airport hotels, baggage
handling, security, and cleaning. However, these spaces on the airport periphery are often
ignored except on the occasion of major aviation disasters. The tragedies of the Bijlmeer

in Amsterdam, a housing estate partly decimated in 1992 by an El Al Boeing 747; or the July 2000 Concorde crash over Gonesse, adjacent to Paris CDG, bring home the connectedness of these apparently disconnected spaces. They thus are part of the urban logic of the metroworld.

Third, as David Pascoe (2001) describes with great eloquence in *Airspaces*, airports have a huge role in the *representational spaces* of everyday life. Pascoe's detective work at these spaces ranges from the musings of Fleming's James Bond to the films of Jacques Tati, spliced with the modernist enthusiasms of Le Corbusier, the Berlin airlift, the tragedies of the crashes that have punctuated aviation history, and the fears of terrorism and hijack that symbolise Europe's tangled implication in the geopolitics of the 'world outside'. Furthermore, the architecture and engineering of new airports is now regarded with the same importance as cathedral- or palace-building once were. European architects such as Norman Foster, Renzo Piano, Richard Rogers, and Rem Koolhaas now eagerly seek out the chance to build these terminals, which testify to what many see as a contemporary right to a 'citizenship of mobility': the right to travel often, safely, and quickly around the world.

Fourth, airports are *regulatory spaces* of great significance for the control of movement in the European Union. Due to the high levels of safety and security required, they are highly controlled environments. They therefore float free in many ways from their surrounding countries – not for nothing is the airport perhaps the paradigmatic 'non-place' (Augé 1995). Hub airports such as Frankfurt may see every second passenger stay in the airport to transfer. They have no interest in leaving and entering the city, and as such the idea of the 'gateway' is of little relevance. These are spaces of transit and

Figure 24 *Barcelona airport. Donald McNeill.*

movement, temporarily grounding the transnationals of the world in the lounges, bars, restaurants, conference centres, and hotels of the terminal area. For Walters (2002b), the airport is a key territory of European integration based upon the promise of free movement of people, which comes down to the micro-design of glass partitions to prevent the passing of documents, to the signage used to sort and order travellers. Thus the Schengen Executive, aware of the need to speed up the processing and scrutiny of each passenger, identified the need for clarity of signage in airports across the EU:

> The sign chosen is the now familiar one of 'EU' (or 'UE') in a circle of stars, with checkpoints for third-country nationals signalled by 'Non-EU Nationals'. The aim is that by standardising the use of this symbol across all major European airport terminals, arrival and departure areas, passengers will become *habituated* to sorting themselves into the appropriate queues: EU and non-EU.
>
> (Walters 2002b: 107)

Walters then makes a crucial point. Unlike the emphasis placed by Shore (2000) on the importance of flags and passports in constructing European identity, here the flag-as-sign and the passport are technologies that speed movement. This practice, this ability to move quickly across once complex borders, is a central, practical, aim of European integration and will be crucial to the success of the integration project.

And so, the airport, that epitome of non-place, may in fact be quite the opposite, filled with emotion, integrated into its surrounding (usually working class) landscape, as authentic an urban space as any *piazza*, and embodying the contradictions of speed and mobility that define the New European experience.

Conclusion

To end this journey through the New Europe in an unglamorous airport lounge may be somewhat anticlimactic, but it sums up, for me at least, the tension between fixity and mobility that I have been trying to trace. To conclude, I want to make three points.

First, this vision of Europe as a continent covered and even defined by banal modes of rapid transport can perhaps be overplayed. I have been at pains to stress that the mobility aspect of Europe is highly uneven, and that many people are 'place bound'. Yet this would be to ignore the fact that many people are at some point in transit, and that their future willingness or enthusiasm to embrace a European project will be bred from this experience. The traditional notion of civic identity has broken down under mass consumerism — where the idea that, say, the town hall or main square of the European city *intra muros* once provided the focus for citizenship, responsibility to place, and belonging to a territorially defined political community. And relatedly, 'we must also abandon that "indigenous" fantasy of a closed world founded once and for all, long ago ... the fantasy of each ethnic group living in isolation, on its own separate territorial island' (Morley 2000: 174). Where does this leave us in relation to the earlier chapter on cities? No more Glaswegians, Berliners, Mancunians? Or do their essences alter, do new stereotypes emerge? Does the dual drudge of suburban living and non-place communication alter the essential characteristic of what Europe is?

Second, this perhaps lends credence to the theory that the European project will be built upon consumerism, upon the fact that modern Europe has always depended upon resource exploitation *elsewhere*, and that the prosperity of the contemporary continent – vastly uneven, of course, but huge compared with Africa for example – cannot be disentangled from this colonial past. Furthermore, it is clear that despite the attempts of eco-socialists and certain branches of the green movement, the New Europe is built on an ever-increasing – and totally unsustainable – dependency on fossil fuels.

Finally, what does this leave us with in terms of an imaginary? How do we reconcile Paasi's (2001) distinction between 'fast' and 'slow' geographies, between spaces of flows and spaces endowed with meaning over decades, centuries of human activity? In *Roissy Express,* Maspero (1994: 27) takes a whole page to even summarise the 'hordes, armies, gangs, battalions, crusaders, revolting peasants, Protestants, Catholics and rebels during the Fronde who have plundered it; the Romans, Huns, Franks, Normans, Armagnacs, Burgundians, English, Spaniards and raiders from Lorraine, and so on up to the Cossacks and the Prussians' that have criss-crossed the landscape that is now a part of metropolitan Paris. Will our consumer society concrete over these histories, or will the ghostly landscapes of deeply-held identity still haunt the formation of the New Europe?

Bibliography

Agnew, J. (1997), 'The myth of backward Italy in modern Europe', in B. Allen and M. Russo (eds), *Revisioning Italy: National Identity and Global Culture*. Minneapolis: University of Minnesota Press. 23–42.

Aldridge, A. (1995), 'The English as they see others: England revealed in Provence', *Sociological Review* 43. 415–34.

Alibhai-Brown, Y. (2001), 'Islam and Euro-identity: Muslims, diversity and inclusion', in D. Leonard and M. Leonard (eds), *The Pro-European Reader*. Basingstoke: Palgrave. 214–19.

Allen, J., Massey, D. and Cochrane A. (1998), *Rethinking the Region*. London: Routledge.

Amin, A. (ed.) (1994), *Post-Fordism: a Reader*. Oxford: Blackwell.

Amin, A. and Thrift, N. (2002), *Cities: Reimagining the Urban*. Oxford: Polity.

Andall, J. (2002), 'Second-generation attitude? African-Italians in Milan', *Journal of Ethnic and Migration Studies* 28: 3. 389–407.

Anderson, B. (1995), *Imagined Communities: Reflections on the Origin and Spread of Nationalism*. London: Verso.

Anderson, J. (1996), 'The shifting stage of politics: new medieval and postmodern territorialities?', *Environment and Planning D: Society and Space*. 14. 133–53.

Anderson, M. (1996), *Frontiers: Territory and State Formation in the Modern World*. Cambridge: Polity/Oxford: Blackwell.

—— and Bort, E. (eds) (1998), *The Frontiers of Europe*. London: Pinter.

Appadurai, A. (1990), 'Disjuncture and difference in the global cultural economy', *Theory, Culture and Society* 7. 295–310.

—— (1996) *Modernity at Large*. Minneapolis: University of Minnesota Press.

Ardagh, J. (1995), *Ireland and the Irish: Portrait of a Changing Society*. Harmondsworth: Penguin.

—— (2000), *France in the New Century: Portrait of a Changing Society*. Harmondsworth: Penguin.

Augé, M. (1995), *Non-places: Introduction to an Anthropology of Supermodernity*. Tr. John Howe. London: Verso.

Baart, T., Metz, T. and Ruimschotel, T. (2000), *Atlas of Change: Rearranging the Netherlands*. Rotterdam: Nai/Ideas on Paper.

Baeten, G. (2001), 'The Europeanization of Brussels and the urbanization of 'Europe': Hybridizing the city. Empowerment and disempowerment in the EU district', *European Urban and Regional Studies* 8. 117–30.

Balkir, C. and Williams, A. M. (1993), *Turkey and Europe*. London: Pinter.

Ballard, J. G. (2001), 'Welcome to the virtual city', *Tate*. 24. Spring. 33.

Basilico, G. and S. Boeri, (1999), *Italy: Cross Sections of a Country.* Zurich: Scalo.

Bayley, S. (1998), *Labour Camp: the Failure of Style over Substance.* London: Batsford.

Bell, J. (2001), *Carchitecture: Where the Car and the City Collide.* Basel: Birkhäuser.

Bell, D. and Valentine, G. (1997), *Consuming Geographies: We Are Where We Eat.* London: Routledge.

Bellier, I. and Wilson, T. M. (eds) (2000), *An Anthropology of the European Union: Building, Imagining and Experiencing the New Europe.* Oxford: Berg.

Benevolo, L. (1993), *The European City.* Trans. Carl Ipsen. Oxford: Blackwell.

Berman, M. (1982), *All that is Solid Melts into Air: the Experience of Modernity.* London: Verso.

Bertolini, L. (2000), 'Planning in the borderless city: a conceptualisation and an application to the case of station area redevelopment', *Town Planning Review* 71. 4. 455–75.

Bialasiewicz, L. and O'Loughlin, J. (2002), 'Galician identities and political cartographies on the Polish-Ukrainian border', in D.H. Kaplan, and J. Häkli (eds) *Boundaries and Place: European Borderlands in Geopolitical Context.* Lanham: Rowman and Littlefield. 217–38.

Bigo, D. (1998), 'Frontiers and security in the European Union: the illusion of migration control', in M. Anderson and E. Bort (eds) *The Frontiers of Europe.* London: Pinter. 148–64.

Billig, M. (1995), *Banal Nationalism.* London: Sage.

Bishop, P. (2002), 'Gathering the land: the Alice Springs to Darwin rail corridor', *Environment and Planning D: Society and Space* 20. 295–317.

Blain, N., Boyle, R. and O'Donnell, H. (1993), *Sport and National Identity in the European Media.* Leicester: Leicester University Press.

Boeri, S. (1999), 'The Italian landscape: towards an "eclectic atlas"', in G. Basilico and S. Boeri, *Italy: Cross Sections of a Country.* Zurich: Scalo. 9–24.

Bogdanovič, B. (1995), 'The city and death', in J. Labon (ed.), *Balkan Blues: Writing out of Yugoslavia.* Evanston, Ill.: Northwestern University Press.

Boomkens, R. (1999) 'The "middle landscape" and the myth of mobility: coming home in commuter country', in Ghent Urban Studies Team, (eds), *The Urban Condition: Space, Community and Self in the Contemporary Metropolis.* Rotterdam: 010. 214–27.

Borneman, J. (1998), '*Grenzregime* (border regime): the Wall and its aftermath', in T. M. Wilson and H. Donnan (eds), *Border Identities: Nation and State at International Frontiers.* Cambridge: Cambridge University Press. 162–90.

Bort, E. (ed.) (1998a), *Borders and Borderlands in Europe.* Edinburgh: International Social Sciences Institute.

—— (1998b), 'Mitteleuropa: the difficult frontier', in M. Anderson and E. Bort (eds) *The Frontiers of Europe.* London: Pinter. 91–108.

Boyle, M. and Hughes, G. (1991), 'The politics of the representation of the "real": discourses from the Left on Glasgow's role as City of Culture, 1990', *Area* 23. 217–28.

Bradbury, M. (1991), 'The Cities of Modernism', in M. Bradbury and J. McFarlane (eds), *Modernism: a Guide to European Literature, 1890–1930.* London: Penguin. 96–104.

—— (1996), *Dangerous Pilgrimages: Trans-Atlantic Mythologies and the Novel.* Harmondsworth: Penguin.

Braudel, F. (1976/1995), *The Mediterranean and the Mediterranean World in the Age of Phillip II.* Berkeley/Los Angeles CA: University of California Press.

Budge, I. and Newton, K. (eds) (1997), *The Politics of the New Europe: Atlantic to Urals.* London: Longman.

Bull, H. (1977), *The Anarchical Society: a Study of Order in World Politics.* London: Macmillan.

Buller, H. and Hoggart, K. (1995), *International Counterurbanisation: British Migrants in Rural France.* Aldershot: Avebury.

Burgess, A. (1997), *Divided Europe: the New Domination of the East.* London: Pluto.

Burns, J. (1999), *Barça: A People's Passion.* London: Bloomsbury.

Çağlar, A. S. (2001), 'Constraining metaphors and the transnationalisation of spaces in Berlin', *Journal of Ethnic and Migration Studies* 27:4. 601–13.

Cameron, F. (2002), 'The Islamic factor in the European Union's foreign policy', in S. T. Hunter (ed.), *Islam, Europe's Second Religion: the new Social, Cultural and Political Landscape.* Westport: Praeger. 257–70.

Campbell, S. (1999), 'Capital reconstruction and capital accumulation in Berlin: a reply to Peter Marcuse', *International Journal of Urban and Regional Research* 23:1. 173–9.

Carr, G. and Paul, G. (1995), 'Unification and its Aftermath: the challenge of history', in R. Burns (ed.), *German Cultural Studies: an Introduction.* Oxford: Oxford University Press. 325–48.

Castells, M. (1994), 'European cities, the informational society, and the global economy', *New Left Review* 204. March/April. 18–32.

—— (1996), *The Rise of the Network Society.* Oxford: Blackwell.

—— (1997a), *End of Millennium.* Oxford: Blackwell.

—— (1997b), *The Power of Identity.* Oxford: Blackwell.

Castells, M. and Hall, P. (1994), *Technopoles of the World: The Making of 21ˢᵗ Century Industrial Complexes.* London: Routledge.

Castles, S. and Miller, M. J. (1998), *The Age of Migration: International Population Movements in the Modern World.* Basingstoke: Macmillan. 2nd edition.

Cembalest, R. (1997). 'First we take Bilbao', *Artforum International.* September, 63–4.

Chalaby, J. K. (2002), 'Transnational television in Europe: the role of pan-European channels', *European Journal of Communication* 17 (2). 183–203.

Chevalier, L. (1994), *The Assassination of Paris.* Chicago: University of Chicago Press.

Chidester, D. (2001), *Christianity: A Global History.* London: Penguin.

Church, A. and Reid, P. (1996), 'Urban power, international networks and competition: the example of cross-border cooperation', *Urban Studies* 33 (8). 1297–1318.

Cobb, R. (1998), *Paris and Other Places.* London: John Murray.

Collard, S. (1992), 'Mission impossible: *les chantiers du President*', *French Cultural Studies*, 2. 97–132.

Collard, S. (1996), 'Politics, culture and urban transformation in Jacques Chirac's Paris 1977–1995', *French Cultural Studies* 7. 1–31.

Commission for the European Communities (1991), *Europe 2000: Outlook for the Development of the Community's Territory.* Luxembourg: Office for Official Publications of the European Communities.

Commission for the European Communities (1994), *Europe 2000+: Co-operation for European Territorial Development.* Luxembourg: Office for Official Publications of the European Communities.

Commission for the European Communities (1999), *European Spatial Development Perspective.* Luxembourg: Office for Official Publications of the European Communities.

Commission for the European Communities (2001), *White Paper: European Transport Policy for 2010: Time To Decide.* Luxembourg: Office for Official Publications of the European Communities.

Commission for the European Communities (2002), *Trans-European Transport Network: Ten-T Priority Projects.* Luxembourg: Office for Official Publications of the European Communities.

Crang, P. and Jackson P. (2001), 'Geographies of Consumption', in D. Morley and K. Robins (eds), *British Cultural Studies: Geography, Nationality and Identity.* Oxford: Oxford University Press. 327–42.

Crolley, L. and Hand, D. (2002), *Football, Europe and the Press.* London: Frank Cass.

Crystal, D. (1997), *English as a Global Language.* Cambridge: Cambridge University Press.

Darian-Smith, E. (1999), *Bridging Divides: The Channel Tunnel and English Legal Identity in the New Europe.* Berkeley: University of California Press.

Davidson, R. (1998), *Travel and Tourism in Europe.* Harlow: Longman. Second edition.

Davies, N. (1996), *Europe: A History.* Oxford: Oxford University Press.

Delanty, G. (1995), *Inventing Europe: Idea, Identity, Reality.* Basingstoke: Macmillan.

—— (2000), 'The resurgence of the city in Europe? The spaces of European citizenship', in E.F. Isin (ed.). *Democracy, Citizenship and the Global City.* London: Routledge. 79–92.

Dickie, J. (1996), 'Imagined Italies', in D. Forgacs and R. Lumley (eds), *Italian Cultural Studies: an Introduction.* Oxford: Oxford University Press. 19–33.

Downing, D. (2000), *The Best of Enemies: England v. Germany, A Century of Football Rivalry.* London: Bloomsbury.

Drakulič, S. (1987), *How We Survived Communism and Even Laughed.* London: Vintage.

—— (1996), *Cafe Europa.* London: Abacus.

Driessen, H. (1998), 'The "new immigration" and the transformation of the European-African frontier', in T. M. Wilson and H. Donnan (eds), *Border Identities: Nation and State at International Frontiers.* Cambridge: Cambridge University Press. 96–116.

Driver, F. and Gilbert, D. (1999), *Imperial Cities.* Manchester: Manchester University Press.

Driver, S. and Martell, L. (2001), 'Blair and "Britishness"', in D. Morley and K. Robins (eds), *British Cultural Studies: Geography, Nationality and Identity.* Oxford: Oxford University Press. 461–72.

Dunford, M. and Kafkalas, G. (1992), *Cities and Regions in the New Europe: the Global-Local Interplay and Spatial Development Strategies.* London: Belhaven.

Dyson, K. (1994), *Elusive Union: The Process of Economic and Monetary Union in Europe.* Harlow: Longman.

Edensor, T. (1997), 'National identity and the politics of memory: remembering Bruce and Wallace in symbolic space', *Environment and Planning D: Society and Space.* 29. 175–94.

—— (2002), *National Identity, Popular Culture and Everyday Life.* Oxford: Berg.

Eder, S. and Sandtner, M. (2002), 'Common spirit in the Upper Rhine Valley?', in D. H. Kaplan and J. Häkli (eds), *Boundaries and Place: European borderlands in Geopolitical Context.* Lanham: Rowman and Littlefield. 141–58.

Ellwood, D. W. (2000), 'Comparative anti-Americanism in Western Europe', in H. Fehrenbach and U. G. Poiger (eds), *Transactions, Transgressions, Transformations: American Culture in Western Europe and Japan.* New York: Bergbahn 26–44.

Enzensberger, H. M. (1989), *Europe, Europe: Forays into a Continent.* London: Hutchinson.

Espadaler, A. M. (2002), '¿Es la Rambla un paseo?', *La Vanguardia (Digital)* 7 July. www.lavanguardia.es

Esposito, J. L. (2002), 'The Muslim diaspora and the Islamic world', in S. T. Hunter (ed.), *Islam, Europe's Second Religion: The New Social, Cultural and Political Landscape.* Westport: Praeger. 245–55.

Evans, G. (2001), *Cultural Planning: An Urban Renaissance.* London: Routledge.

Evans, M. (1996), 'Languages of racism within contemporary Europe', in B. Jenkins and S. A. Sofos (eds), *Nation and Identity in Contemporary Europe.* London: Routledge.

Fantasia, R. (1995), 'Fast food in France', *Theory and Society* 24. 201–43.

Fehrenbach, H. and Poiger, U.G. (2000), 'Introduction: Americanization reconsidered', in H. Fehrenbach and U. G. Poiger (eds), *Transactions, Transgressions, Transformations: American culture in Western Europe and Japan.* New York: Bergbahn xiii–xl.

Foot, J. (1999), 'From boomtown to bribesville: the images of the city, Milan 1980–97'. *Urban History* 26, 3. 393–412.

—— (2000), 'The urban periphery, myth and reality: Milan, 1950–2000', *City* 4, 1. 7–26.

—— (2001), *Milan since the Miracle: City, Culture and Identity.* Oxford: Berg.

Forest, B. and Johnson, J. (2002), 'Unraveling the threads of history: Soviet-era monuments and post-Soviet national identity in Moscow', *Annals of the Association of American Geography* 92 (3), 524–47.

Foucher, M. (1998), 'The geopolitics of European frontiers', in M. Anderson and E. Bort (eds), *The Frontiers of Europe.* 235–50.

Fraser, N. (1998) *Continental Drifts: Travels in the New Europe.* London: Vintage.

—— (2000), *The Voice of Modern Hatred: Encounters with Europe's New Right.* London: Picador.

Freemantle, B. (1995), *The Octopus: Europe in the Grip of Organized Crime.* London: Orion.

Frisby, D. (2001), *Cityscapes of Modernity: Critical Explorations.* Cambridge: Polity.

Fyfe, N. (1998), *Images of the Street: Planning, Identity and Control in Public Space.* London: Routledge.

Garreau, J. (1991), *Edge City: Life on the New Frontier.* New York: Anchor.

Garton Ash, T. (1990), *We the People: the Revolution of '89, Witnessed in Warsaw, Budapest, Berlin and Prague.* Cambridge: Granta.

—— (2000), *History of the Present: Essays, Sketches, and Despatches from Europe in the 1990s.* Harmondsworth: Penguin.

—— (2001), 'The European orchestra', *New York Review of Books.* 17 May. 60–67.

Ginsborg, P. (2001), *Italy and its Discontents: Family, Civil Society, State 1980–2001.* London: Allen Lane/Penguin.

Giordano, B. (2000), 'Italian regionalism or 'Padanian' nationalism – the political project of the Lega Nord in Italian politics', *Political Geography* 19. 445–71.

Goodman, S. (1996), 'Market forces speak English', in S. Goodman and D. Graddol (eds) (1996), *Redesigning English: New Texts, New Identities*. London: Routledge/Open University. 141–80.

Graddol, D. (1996), 'Global English, global culture?', in Goodman, S. and D. Graddol (eds) (1996), *Redesigning English: New Texts, New Identities*. London: Routledge/Open University. 181–238.

Graham, B. (1997), 'Ireland and Irishness: Place, culture and identity', in B. Graham (ed.), *In Search of Ireland: a Cultural Geography*. London: Routledge. 1–15.

Graham, H. and Sánchez, A. (1995), 'The politics of 1992', in H. Graham and J. Labanyi (eds), *Spanish Cultural Studies: An Introduction*. Oxford: Oxford University Press. 406–18.

Graham, S. and Marvin, S. (2001), *Splintering Urbanism: Networked Infrastructures, Technological Mobilities and the Urban Condition*. London: Routledge.

Gregory, D. (1994), *Geographical Imaginations*. Oxford: Blackwell.

Grodach, C. (2002), 'Reconstituting identity and history in post-war Mostar, Bosnia-Herzegovina', *City* 6 (1). 61–82.

Guerrina, R. (2002), *Europe: History, Ideas and Ideologies*. London: Arnold.

GUST (Ghent Urban Studies Team) (1999), *The Urban Condition: Space, Community and Self in the Contemporary Metropolis*. Rotterdam: 010.

Hadjimichalis, C. and Sadler, D. (1995), *Europe at the Margins: New Mosaics of Inequality*. Chichester: John Wiley.

Hagen, J. (2003), 'Redrawing the imagined map of Europe: the rise and fall of the "center"', *Political Geography* 22. 489–517.

Häkli, J. (2002), 'Transboundary networking in Catalonia', in D. H. Kaplan, and J. Häkli (eds) *Boundaries and Place: European Borderlands in Geopolitical Context*. Lanham: Rowman and Littlefield. 70–92.

Halewood, C. and Hannam, K. (2001), 'Viking heritage tourism: authenticity and commodification', *Annals of Tourism Research* 28. 565–80.

Hall, P. (1999), *Cities in Civilisation*. London: Weidenfeld and Nicolson.

Hall, P. (2002), 'Europe and cities', in D. Leonard and M. Leonard (eds), *The Pro-European Reader*. Basingstoke: Palgrave/Foreign Policy Centre. 196–203.

Hall, S. (1995), 'New cultures for old', in D. Massey and P. Jess (eds), *A Place in the World: Places, Cultures and Globalization*. Oxford: OUP/Open University.

Hall, T. and Hubbard, P. (1996), 'The entrepreneurial city: new urban politics, new urban geographies?', *Progress in Human Geography* 20. 153–74.

—— (1998), *The Entrepreneurial City: Geography of Politics, Regime and Representation*. Chichester: John Wiley.

Hannerz, U. (1996), *Transnational Connections: Culture, People, Places*. London: Routledge.

Hargreaves, J. (2000), *Freedom for Catalonia? Catalan Nationalism, Spanish Identity, and the Barcelona Olympic Games*. Cambridge: Cambridge University Press.

Harvey, D. (1989), 'From managerialism to entrepreneurialism: the transformation of governance in late capitalism', *Geografiska Annaler*, 71B. 3–17.

Harvie, C. (1994), *The Rise of Regional Europe*. London: Routledge.

Haslam, D. (1999), *Manchester England: The Story of the Pop Cult City*. London: Fourth Estate.

Häussermann, H. (1999), 'Economic and political power in the New Berlin': a response to Peter Marcuse', *International Journal of Urban and Regional Research* 23:1. 180–84.

Hearn, P. M. and Button, D. E. (eds) (1994), *Language Industries Atlas*. Amsterdam: IOS.

Heffernan, M. (1998), *The Meaning of Europe: Geography and Geopolitics*. London: Arnold.

Hertmans, S. (2001), *Intercities*. London: Reaktion.

Hobsbawm, E. and Ranger, T. (1992), *The Invention of Tradition*. Cambridge: Cambridge University Press.

Hooper, J. (1995), *The New Spaniards*. Harmondsworth: Penguin.

Hoyle, B. S. (ed.) (1996), *Cityports, Coastal Zones and Regional Change: International Perspectives on Planning and Management*. Chichester: John Wiley.

—— and Pinder, D. A. (1992a), 'Cities and the sea: change and development in contemporary Europe', in B. S. Hoyle and D. A. Pinder (eds), *European Port Cities in Transition*. London: Belhaven/British Association for the Advancement of Science. 1–19.

—— and Pinder, D. A. (eds), (1992b), *European Port Cities in Transition*. London: Belhaven/British Association for the Advancement of Science.

Hughes, G. (1998), 'Tourism and the semiological realization of space', *Destinations: Cultural Landscapes of Tourism*. London: Routledge. 17–32.

Hutchinson, J. (2003), 'Enduring nations and the illusions of European integration', in W. Spohn and A. Triandafyllidou (eds), *Europeanisation, National Identities and Migration: Changes in Boundary Constructions between Western and Eastern Europe*. London: Routledge. 36–51.

Isin, E. F. and Wood, P. K. (1999), *Citizenship and Identity*. London: Sage.

Jacobs, J. M. (1994), 'Negotiating the heart: heritage, development and identity in postimperial London', *Environment and Planning D: Society and Space*. 12. 751–72.

Jacobs, J. (1996), *Edge of Empire: Postcolonialism and the City*. London: Routledge.

Jansen, S. (2001), 'The streets of Beograd. Urban space and protest identities in Serbia', *Political Geography* 20. 35–55.

Jaureguí, P. (1999), 'National pride and the meaning of "Europe": a comparative study of Britain and Spain', in D. Smith and S. Wright (eds), *Whose Europe? The Turn Towards Democracy*. Oxford: Blackwell/Sociological Review. 257–87.

Jensen, O. B. and Richardson, T. (2003), 'Being on the map: the new iconographies of power over European space', *International Planning Studies* 8. 9–34.

Jones, J. (2000), 'Come, friendly bombs…', The *Guardian*, 8 January. Review p. 5.

Jönsson, C., Tägli, S. and Törnqvist, G. (2000), *Organising European Space*. London: Sage.

Judd, D. R. (1999), 'Constructing the tourist bubble', in D. R. Judd and S. S. Fainstein (eds). *The Tourist City*. New Haven: Yale University Press. 35–53.

Kaplan, D. H. (2002), 'Asymmetrical and hybrid identities in the Northern Italian borderlands', in D. H. Kaplan, and J. Häkli (eds) *Boundaries and Place: European Borderlands in Geopolitical Context*. Lanham: Rowman and Littlefield. 116–40.

—— and J. Häkli (eds) (2002), *Boundaries and Place: European Borderlands in Geopolitical Context*. Lanham: Rowman and Littlefield.

Karppi, K. (2002), 'Symbolic and functional balance on Europe's Northern borders', in D. H. Kaplan, and J. Häkli (eds) *Boundaries and Place: European Borderlands in Geopolitical Context*. Lanham: Rowman and Littlefield. 255–72.

Kearney, R. (1995), 'Culture: the price you pay', interview with George Steiner, in R. Kearney (1995), *Dialogues with Contemporary Thinkers on the European Mind*. Manchester: Manchester University Press. 82–92.

Kearns, G. (1993), 'The city as spectacle: Paris and the bicentenary of the French Revolution', in G. Kearns and C. Philo (eds), *Selling Places: the City as Cultural Capital, Past and Present*. Oxford: Pergamon. 49–101.

Keating, M. (1998), *The New Regionalism in Western Europe: Territorial Restructuring and Political Change*. Cheltenham: Edward Elgar.

Keil, R. (1994), 'Global sprawl: urban form after Fordism', *Environment and Planning D: Society and Space* 12. 131–6.

—— (1999), 'Germany: a middle landscape in winter', *Environment and Planning D: Society and Space* 17. 632–9.

—— and Ronneberger, K. (1994), 'Going up in the country: internationalization and urbanization on Frankfurt's northern fringe', *Environment and Planning D: Society and Space*. 12. 137–66.

Keiller, P. (1999), *Robinson in Space*. London: Reaktion.

—— (2001), 'Port statistics', in I. Borden, J. Kerr and J. Rendell (eds), *The Unknown City: Contesting Architecture and Social Space*. Cambridge MA: MIT Press. 443–57.

Kelly-Holmes, H. (2000), 'Bier, parfum, kaas: language fetish in European advertising', *European Journal of Cultural Studies* 5 (1). 67–82.

Kepka, J. M. M. and Murphy, A. B. (2002), 'Euroregions in comparative perspective', in D. H. Kaplan and J. Häkli (eds) *Boundaries and Place: European Borderlands in Geopolitical Context*. Lanham: Rowman and Littlefield. 50–69.

Kiberd, D. (2002), 'The city in Irish culture', *City* 6 (2), 219–28.

King, R. (1997), 'Migrations, globalization and place', in D. Massey and P. Jess (eds), *A Place in the World? Places, Culture and Globalization*. Oxford: Oxford University Press. 5–44.

—— (1998), 'The Mediterranean: Europe's Rio Grande', in M. Anderson and E. Bort (eds) *The Frontiers of Europe*. London: Pinter. 109–34.

Klemencic, M. and Schofield, C. (2002), in D. H. Kaplan, and J. Häkli (eds) *Boundaries and Place: European Borderlands in Geopolitical Context*. Lanham: Rowman and Littlefield. 200–16.

Kobrin, S. J. (1998), 'Back to the future: neomedievalism and the postmodern digital world economy', *Journal of International Affairs*. 51. 361–86.

Koff, S. Z. and Koff, S. P. (2000), *Italy: From the First to the Second Republic*. London: Routledge.

Koolhaas, R. (1996) 'Beyond delirious', in K. Nesbitt (ed.), *Theorising a New Agenda for Architecture: An Anthology of Architectural Theory. 1965–1995*. New York: Princeton Architectural Press. 332–6.

Kramsch, O. T. (2002), 'Reimagining the scalar topographies of cross-border governance: Eu(ro)regions in the post-colonial present', *Space and Polity* 6 (2) 169–96.

Kuisel, R. (1993), *Seducing the French: the Dilemma of Americanization.* Berkeley: University of California.

Kunzmann, K. (1996), 'Euro-megalopolis or Themepark Europe? Scenarios for European Spatial Development', *International Planning Studies* 1 (2), 143–63.

Kuper, S. (2002), 'Football and Europe', in D. Leonard and M. Leonard (eds), *The Pro-European Reader.* Basingstoke: Palgrave/Foreign Policy Centre. 204–13.

Kürti, L. (1997), 'Globalisation and the discourse of otherness in the 'New' Eastern and Central Europe' in T. Modood and P. Werbner (eds), *The Politics of Multiculturalism in the New Europe: Racism, identity and community.* London: Zed. 29–53.

Ladd, B. (1997), *Ghosts of Berlin: Confronting German History in the Urban Landscape.* Chicago: University of Chicago Press.

Laermans, R. (1999), 'Downtown Brussels: a portrait', in Ghent Urban Studies Team (eds), *The Urban Condition: Space, Community and Self in the Contemporary Metropolis.* Rotterdam: 010 Publishers. 300–07.

Lainsbury, A. (2000), *Once Upon an American Dream: the Story of Euro Disneyland.* Lawrence KA: University Press of Kansas.

Lanfranchi, P., and Taylor, M. (2001). *Moving with the Ball: The Migration of Professional Footballers.* Oxford: Berg.

Leach, N. (ed.) (1999), *Architecture and Revolution: Contemporary Perspectives on Central and Eastern Europe.* London: Routledge.

Le Galès, P. (2002), *European Cities: Social Conflict and Governance.* Oxford: Oxford University Press.

Lees-Milne, J. (1967), *Saint Peter's: The Story of Saint Peter's Basilica in Rome.* London: Hamish Hamilton.

Lennon, D. and Foley, M. (2000), *Dark Tourism: The Attraction of Death and Disaster.* London: Continuum.

Lentin, R. (2002), 'At the heart of the Hibernian post-metropolis: spatial narratives of ethnic minorities and diasporic communities in a changing city', *City* 6 (2). 229–49.

Leonard, D. and Leonard, M. (eds) (2001), *The Pro-European Reader.* Basingstoke: Palgrave.

Leung, M. W. H. (2001), 'Get IT going: new ethnic Chinese business. The case of Taiwanese-owned computer firms in Hamburg', *Journal of Ethnic and Migration Studies* 27: 2. 277–94.

Looseley, D. L. (1995), *The Politics of Fun: Cultural Policy and Debate in Contemporary France.* Oxford: Berg.

Lupo, S. (1996), 'The changing Mezzogiorno: between representations and reality', in S. Gundle and S. Parker (eds), *The New Italian Republic: From the Fall of the Berlin Wall to Berlusconi.* London: Routledge. 130–46.

MacDonald, F. (2000), *The Construction of Dublin.* Kinsale: Gandon Editions.

McCarthy, P. (1996), 'Forza Italia: the new politics and old values of a changing Italy', in S. Gundle and S. Parker (eds), *The New Italian Republic: From the Fall of the Berlin Wall to Berlusconi.* London: Routledge. 130–46.

McGill, C. (2001), *Football Inc.: How Soccer Fans are Losing the Game.* London: Vision.

McNeill, D. (1999a), 'Globalisation and the European city', *Cities.* 16, 3. 143–7.

—— (1999b) *Urban Change and the European Left: Tales from the New Barcelona*. London: Routledge.

—— (2000), 'McGuggenisation: globalisation and national identity in the Basque country', *Political Geography*. 19. 473–94.

—— (2001a), 'Barcelona as imagined community: Pasqual Maragall's spaces of engagement', *Transactions of the Institute of British Geographers* 26. 340–52.

—— (2001b) 'Embodying a Europe of the cities: the geographies of mayoral leadership', *Area* 33, 353–9.

—— (2001c), 'Rutelli's Rome: mayors and personality politics in Italy', *European Urban and Regional Studies* 8. 253–6.

—— (2003), 'Rome, global city? Church, state and the Jubilee 2000', *Political Geography* 22. 535–56.

Magris, C. (1990), *Danube: A Sentimental Journey from the Source to the Black Sea*. Trans. Patrick Creagh. London: Collins Harvill.

Maguire, M. and Hollywood, P. (2002), 'Introduction: the city in the era of globalization', *City* 6 (2),167–72.

Mann, M. (1998), 'Is there a society called Euro?', in R. Axtmann (ed.), *Globalization and Europe: Theoretical and Empirical Investigations*. London: Pinter. 184–207.

Marcuse, P. (1998) 'Reflections on Berlin: the meaning of construction and the construction of meaning', *International Journal of Urban and Regional Research* 22 (2). 331–8.

—— (1999) 'Reply to Campbell and Häussermann', *International Journal of Urban and Regional Research* 23:1. 185–7.

Markwick, M. (2001), 'Marketing myths and the commodification of Ireland', *Geography*. 86. 37–49.

Marks, J. (1999), 'The French national team and national identity: 'Cette France d'un "bleu métis"', in H. Dauncey and G. Hare (eds), *France and the 1998 World Cup: The National Impact of a World Sporting Event*. London: Frank Cass. 41–57.

Maspero, F. (1994), *Roissy Express: A Journey Through the Paris Suburbs*. London: Verso.

Massey, D. (1991), 'A global sense of place', *Marxism Today* June. 24–9.

—— (1995), 'Places and their pasts', *History Workshop Journal* 39.182–92.

—— (1999), 'Cities in the world', in D. Massey, J. Allen and S. Pile (eds), *City Worlds*. London: Routledge. 99–136.

—— Quintas, P. and Wield, D. (1992), *Hi-Tech Fantasies: Science Parks in Society, Science and Space*. London: Routledge.

Mayle, P. (1990), *A Year in Provence*. London: Pan.

Michael, M. (2001), 'The invisible car: the cultural purification of road rage', in D. Miller (ed.), *Car Cultures*. Oxford: Berg. 59–80.

Michalski, S. (1998), *Public Monuments: Art in Political Bondage 1870–1997*. London: Reaktion Books.

Middleton, C. (1995), 'It's all the rage of the road'. *Sunday Times*, 25 June.

Millward, A. (1992), *The European Rescue of the Nation-State*. Berkeley/Los Angeles: University of California Press.

Miyoshi, M. (2002), 'The university in the "global" economy', in K. Robins and F. Webster (eds), *The Virtual University? Knowledge, Markets and Management.* Oxford: Oxford University Press 49–83.

Moretti, F. (1999), *Atlas of the European Novel, 1800–1900.* London: Verso.

Morgan, S. J. (1999), 'The ghost in the luggage: Wallace and Braveheart: post-colonial "pioneer" identities', *European Journal of Cultural Studies* 2 (3). 375–92.

Moriconi-Ebrard, F. (2000), *De Babylone à Tokyo: les Grandes Agglomérations du Monde.* Paris: Ophrys.

Morley, D. (2000), *Home Territories: Media, Mobility and Identity.* London: Routledge.

—— (2001), 'The national culture in its new global context', in D. Morley and K. Robins (eds), *British Cultural Studies: Geography, Nationality and Identity.* Oxford: Oxford University Press. 1–15.

—— and Robins, K. (1995), *Spaces of Identity: Global Media, Electronic Landscapes, and Cultural Boundaries.* London: Routledge.

Morris, J. (1998), *Fifty Years of Europe: An Album.* London: Penguin.

Morrison, B. (2001), 'Kafka's castle', *Guardian* 18 December. (review p. 7).

Murphy-Lejeune, E. (2002), *Student Mobility and Narrative in Europe: The New Strangers.* London: Routledge.

Nash, C. (1993), '"Embodying the nation": the West of Ireland landscape and national identity', in M. Cronin and B. O'Connor (eds), *Tourism and Ireland.* Cork: Cork University Press. 86–114.

Neill, W. V. (1997), 'Memory, collective identity and urban design: the future of Berlin's Palast der republik', *Journal of Urban Design* 2: 2. 179–92.

Neutelings, W.-J. (1988), *De ringcultuur.* Mechelen: Vlees and Beton.

—— (1994), 'Explorations into wonderland: riding the periphery of the Low Countries', in J. Woodroffe, D. Papa, and I. MacBurnie (eds), 'The Periphery'. *Architectural Design.* 108. London: Wiley. 58–60.

New Statesman and Society (1992), 'Gazetteer', in 'Borderlands' supplement, 19 June. 16–29.

Newman, D. and Paasi, A. (1998), 'Fences and neighbours in the postmodern world: boundary narratives in political geography', *Progress in Human Geography* 22 (2). 186–207.

Newman, P. and Thornley, A. (1996), *Urban Planning in Europe: International Competition, National Systems and Planning Projects.* London: Routledge.

Nicolson, A. (1999), *Regeneration: the Story of the Dome.* London: HarperCollins.

Nugent, N. (2002), *The Government and Politics of the European Union.* Basingstoke: Palgrave Macmillan.

O'Byrne, D. (2001), 'On passports and border controls', *Annals of Tourism Research* 28, 2. 399–416.

O'Dowd, L. (2002), 'Transnational integration and cross-border regions in the European Union', in J.Anderson (ed.), *Transnational Democracy: Political Spaces and Border Crossings.* London: Routledge. 111–28.

—— and Wilson, T. M. (1996), *Borders, Nations and States: Frontiers of Sovereignty in the New Europe.* Aldershot: Avebury.

O'Reilly, K. (2000), *The British on the Costa del Sol: Transnational Identities and Local Communities*. London: Routledge.

O'Toole, F. (1994), *Black Hole, Green Card: The Disappearance of Ireland*. Dublin: New Island Books.

Orwell, G. (1938/1989) *Homage to Catalonia*, Harmondsworth, Penguin.

Paasi, A. (2001), 'Europe as a social process and discourse: considerations of place, boundaries and identity', *European Urban and Regional Studies* 8. 7–28.

—— (2002), 'Place, boundaries and the construction of Finnish territory', in D. H. Kaplan, and J. Häkli (eds) *Boundaries and Place: European Borderlands in Geopolitical Context*. Lanham: Rowman and Littlefield. 178–99.

Parissien, S. (2001), *Station to Station*. London: Phaidon.

Pascoe, D. (2001), *Airspaces*. London: Reaktion.

Pawley, M. (1998), *Terminal Architecture*. London: Reaktion.

Pells, R. (1997), *Not Like Us: How Europeans have Loved, Hated, and Transformed American Culture since World War II*. New York: Basic Books.

Phillips, C. (1987), *The European Tribe*. London: Faber.

Pile, S. (1999), 'What is a city', in D. Massey, J. Allen and S. Pile (eds), *City Worlds*. London: Routledge. 3–52.

Pile, S. and Thrift, N. (eds) (2000), *City A-Z*. London: Routledge.

Pinder, D. (ed.) (1998), *The New Europe: Economy, Society and Environment*. Chichester: John Wiley.

Plant, M. (2003), *Venice, Fragile City: 1797–1997*. New Haven: Yale University Press.

Pointon, M. (1998), 'Money and nationalism', in G. Cubitt (ed.), *Imagining Nations*. Manchester: Manchester University Press. 229–54.

Pollard, J. and Sidaway, J. D. (2002), 'Editorial: Euroland – economic, cultural and political geographies', *Transactions of the Institute of British Geographers* 27. 7–10.

Pred, A. (1995), *Recognising European Modernities: A Montage of the Present*. London: Routledge.

—— (2000), *Even in Sweden: Racisms, Racialized Spaces, and the Popular Geographic Imagination*. Berkeley: University of California Press.

Raento, P. (2002), 'Integration and division in the Basque borderland', in D. H. Kaplan, and J. Häkli (eds) *Boundaries and Place: European Borderlands in Geopolitical Context*. Lanham: Rowman and Littlefield. 93–115.

Rawnsley, A. (2001), *Servants of the People: The Inside Story of New Labour*. Harmondsworth: Penguin.

Richardson, T. and Jensen, O. B. (2000), 'Discourses of mobility and polycentric development: a contested view of European spatial planning', *European Planning Studies* 8 (4). 503–20.

Richie, A. (1999), *Faust's Metropolis: A History of Berlin*. London: Harvill.

Rietbergen, P. (1998), *Europe: A Cultural History*. London: Routledge.

Robb, P. (1998), *Midnight in Sicily*. London: Harvill.

Roberts, R, and Kynaston, D. (2001), *City State: A Contemporary History of the City of London and how Money Triumphed.* London: Profile.

Robins, K. (1999), 'Europe', in P. Cloke, P. Crang, and M. Goodwin (eds), *Introducing Human Geographies.* London: Arnold. 268–76.

—— (2001), 'To London: the city beyond the nation', in K. Robins and D. Morley (eds), *British Cultural Studies: Geography, Nationality and Identity.* Oxford: Oxford University Press. 473–93.

Roche, M. (2000), *Olympics and Expos in the Growth of Global Culture.* London: Routledge.

Roe, K. and de Meyer, G. (2000), 'Music Television: MTV-Europe', in J. Wieten, G. Murdock, P. Dahlgren (eds), *Television Across Europe.* London: Sage. 141–57.

Ross, J. F. L. (1998), *Linking Europe: Transport Policies and Politics in the European Union.* Westport, Connecticut: Praeger.

Sampson, A. (1968), *The New Europeans.* London: Panther.

Sante, L. (1998) *Factory of Facts.* London: Granta.

Schivelbusch, W. (1987), *The Railway Journey: The Industrialization of Time and Space in the 19th Century.* Berkeley/Los Angeles: University of California Press.

Schlesinger, P. (1997), 'From cultural defence to political culture: media, politics and collective identity in the European Union', *Media, Culture and Society* 19. 369–91.

Schorske, C. (1963), 'The idea of the city in European thought: Voltaire to Spengler', in O. Handlin and J. Burchard (eds.), *The Historian and the City.* Cambridge (USA): MIT Press/Harvard University Press. 95–114.

Sharp, J. P. (1999), 'Writing over the map of Provence: the touristic therapy of A Year in Provence.' In J. Duncan and D. Gregory (eds), *Writes of Passage: Reading Travel Writing.* London: Routledge. 200–18.

Sheller, M. and Urry, J. (2000), 'The city and the car', *International Journal of Urban and Regional Research* 24/4. 737–57.

Shore, C. (2000), *Building Europe: The Cultural Politics of European Integration.* London: Routledge.

Sidaway, J. D. (2001), 'Rebuilding bridges: a critical geopolitics of Iberian transfrontier cooperation in a European context', *Environment and Planning D: Society and Space* 9. 743–78.

Sieverts, T. (1997), *Zwischenstadt: Zwischen Ort und Welt, Raum und Zeit, Stadt und Land.* Wiesbaden: Vieweg.

Sinclair, I. (2002a), 'Hail to the tarmac tourniquet', *Time Out* 1674. 18–25 September. 16–18.

—— (2002b), *London Orbital: A Walk Around the M25.* London: Granta.

Smith, A. (2002), 'Imagining geographies of the "new Europe": geo-economic power and the new European architecture of integration', *Political Geography* 21. 647–70.

Smith, M. P. (2001), *Transnational Urbanism: Locating Globalization.* Malden, MA: Blackwell.

Smith. N. (1996), *The New Urban Frontier: Gentrification and the Revanchist City.* London: Routledge.

Soja, E. W. (2000), *Postmetropolis: Critical Studies of Cities and Regions.* Oxford: Blackwell.

Solnit, R. (2000), *Wanderlust.* London: Verso.

Space and Polity (2002), Special issue on 'The geopolitics of cross-border co-operation in the European Union', Vol. 6, 2.

Sparke, M. (2000), '"Chunnel Visions": unpacking the anticipatory geographies of an anglo-European borderland', *Journal of Borderland Studies* 15. 187–219.

Spohn, W. and Triandafyllidou, A. (2003), *Europeanisation, National Identities and Migration: Changes in Boundary Constructions Between Western and Eastern Europe*. London: Routledge.

Spring, I. (1990), *Phantom Village*. Edinburgh: Polygon.

Stenning, A. (2000), Reviews of Budge and Newton (1997), Pinder (1998), and Unwin (1998), *Transactions of the Institute of British Geographers* 25. 109–11.

Sudjic, D. (1992), *The 100 Mile City*. London: Flamingo.

Swyngedouw, E., Moulaent, F. and Rodriguez, A. (2002), 'Neoliberal urbanization in Europe: large-scale urban development projects and the new urban policy', *Antipode* 34, 3. 542–77.

Tambini, D. (2001), *Nationalism in Italian Politics: the Stories of the Northern League, 1980–2000*. London: Routledge.

Till, K. E. (1999), 'Staging the past: landscape designs, cultural identity, and *erinnerungspolitik* at Berlin's Neue Wache', *Ecumene* 6 (3). 251–83.

Todorova, M. (1997), *Imagining the Balkans*. Oxford: Oxford University Press.

Tomlinson, J. (1999), *Globalization and Culture*. Cambridge: Polity.

Treanor, J. (2000), 'Branding bridges the great divide', *Guardian* 5 July. (www.guardian-unlimited.co.uk/Archive/Article/0,4273,4036893,00.html). Accessed 9 October, 2002.

Unwin, T. (ed.) (1998), *A European Geography*. London: Addison Wesley Longman.

Unwin, T. and Hewitt, V. (2001), 'Banknotes and national identity in central and eastern Europe', *Political Geography* 20. 1005–1028.

Urry, J. (1990), *The Tourist Gaze*. London: Sage.

—— (1995), *Consuming Places*. London: Routledge.

—— (2000), *Sociology Beyond Societies: Mobilities for the Twenty-First Century*. London: Routledge.

Van Ham, P. (2001), *European Integration and the Postmodern Condition: Governance, Democracy, Identity*. London: Routledge.

—— (2002), 'Branding territory: inside the wonderful worlds of PR and IR theory', *Millennium: a Journal of International Studies* 31.2. 249–269.

Vázquez Montalbán, M. (1990), *The Angst-Ridden Executive*. Trans. Ed Emery. London: Serpent's Tail.

—— (1992a), *An Olympic Death*. Trans. E. Emery. London: Serpent's Tail.

—— (1992b), *Barcelonas*. Trans. A. Robinson. London: Verso.

Vertovec, S. (2001), 'Transnationalism and identity', *Journal of Ethnic and Migration Studies* 27:4. 573–82.

Virilio, P. (1991), *The Lost Dimension*. New York: Semiotext(e).

Walker, M. and Travis, A. (1998), 'Losers in the game without frontiers', *Guardian*. 7 January. p. 13.

Walters, W. (2002a), 'Mapping Schengenland: denaturalizing the border', *Environment and Planning D: Society and Space* 20. 561–80.

Walters, W. (2002b), 'The power of inscription: beyond social construction and deconstruction in European Integration Studies', *Millennium: Journal of International Studies* 31 (1). 83–108.

Washington Times (1999), 'Oresund bridge to link two countries' economies', 27 October. (www.internationalspecialreports.com/archives/99/denmark/9.html). Accessed 9 October, 2002.

White, J. (2001), *London in the Twentieth Century: A City and its People*. London: Viking.

Williams, J. (1997), 'The "New Football" in England and Sir John Hall's "New Geordie Nation"', in S. Gehrmann (ed.), *Football and Regional Identity in Europe*. Münster: Lit Verlag. 243–77.

Wilson, E. (1995), 'The rhetoric of urban space', *New Left Review* 209. 146–60.

Wilson, K. (1994), *Channel Tunnel Visions, 1850–1945*. London: Hambledon Press.

Wilson, T. M. and Donnan H. (eds) (1998a), *Border Identities: Nation and State at International Frontiers*. Cambridge: Cambridge University Press.

Wise, M. (2000), 'The Atlantic Arc: transnational European reality or regional mirage?', *Journal of Common Market Studies* 38 (5). 865–90.

Wise, M. Z. (1998), *Capital Dilemma: Germany's Search for a New Architecture of Democracy*. New York: Princeton Architectural Press.

—— (2002), 'A capital of Europe? Brussels is primping' *New York Times* 2 March (www.nyt.com)

Wright, S. (1999), 'A community that can communicate? The linguistic factor in European integration', in D. Smith and S. Wright (eds), *Whose Europe? The Turn Towards Democracy*. Oxford: Blackwell/Sociological Review. 79–103.

—— (2000), *Community and Communication: The Role of Language in Nation State Building and European Integration*. Clevedon: Multilingual Matters.

Yalouri, E. (2001), *The Acropolis: Global Fame, Local Claim*. Oxford: Berg.

Yates, S. (1996), 'English in cyberspace', in S. Goodman, and D. Graddol, (eds), *Redesigning English: New Texts, New Identities*. London: Routledge/Open University.106–40.

Zulaika, J. (1997), *Crónica de una Seducción: El Museo Guggenheim Bilbao*. Nerea, Madrid.

Index